DYING *for* HEAVEN

DYING *for* HEAVEN

Holy Pleasure and Suicide Bombers—Why the Best Qualities of Religion Are Also Its Most Dangerous

ARIEL GLUCKLICH

HarperOne
An Imprint of HarperCollins*Publishers*

HarperCollins books may be purchased for educational, business, or sales promotional use. For information please write: Special Markets Department, HarperCollins Publishers, 10 East 53rd Street, New York, NY 10022.

HarperCollins Web site: http://www.harpercollins.com

FIRST HARPERCOLLINS PAPERBACK EDITION PUBLISHED IN 2010

Library of Congress Cataloging-in-Publication Data

Glucklich, Ariel.
Dying for heaven : holy pleasure and suicide bombers—why the best qualities of religion are also its most dangerous / Ariel Glucklich. — 1st ed.
p. cm.
Includes index.
ISBN 978-0-06-143082-4
1. Terrorism—Religious aspects—Islam. 2. Suicide bombers—Religious life. 3. Terrorists—Religious life. 4. Violence—Religious aspects. 5. Hedonism—Religious aspects. 6. Heaven. 7. Paradise. 8. Future life. I. Title.
BP190.5.T47G57 2009
201'.76332—dc22 2009005174

10 11 12 13 14 RRD(H) 10 9 8 7 6 5 4 3 2 1

CONTENTS

Introduction

IMAGINE FINDING YOURSELF IN A game of chicken involving two cars of equal size, both traveling at great speed, where a collision means severe injury or death. Winning will earn you $1,000, which you could use but are not desperate to have. What should you find out about your opponent beforehand? Clearly you would like to know about your opponent's experiences with depression or suicidal fantasies, his family situation, and so forth. And of course, you need to ascertain how much your opponent needs the money. Should you also wonder about his religion? For example, does it make any difference what a player believes about the afterlife, or the value his religion places on courage or on money? Should it matter that the driver in the other car hates your religion or that his minister is watching?

You are trying to predict your opponent's behavior in an extreme situation, and there is a strong possibility that his religious outlook matters. But how do you gauge its impact on a potential head-on collision?

This, roughly speaking, is the situation that many of us could confront in the near future, but on a far grander scale than two cars charging at each other on a country road. Pakistan and India are both armed with nuclear weapons and sophisticated delivery systems. Iran will undoubtedly join them in the coming decade, and Israel is so armed as well. International embargoes, threats of preventive strikes, and even actual air raids are not likely to change the fact that sooner or later the Middle East or South Asia will approach a nuclear abyss. Experts in nuclear deterrence are already hard at work calculating whether deterrence can reduce the risk, as it did during the cold war, given the political, economic, military, and social conditions that prevail in the countries of these regions. Increasingly, they are also looking at religion.

Given everything we know about religion and about the devastating effects of nuclear and biological weapons, can we safely assume that deterrence will work? Can we measure the impact of faith in God on a nuclear showdown or any other major conflict? Israel's recent experiences with both Hizbollah and Hamas raise some serious doubts about the strategy of deterrence. Science can help us predict the rate of glacial melting or the deterioration of bridges on interstate highways. Science and logic can even anticipate the moves of a chess grand master deep into a match, or the response of the Soviet Union to nuclear deterrence. In sharp contrast, religion and one of its chief components, faith in God, are seldom subjected to scientific scrutiny. We avoid scientific evaluation of religion for a variety of reasons, chief among them being the notion that religion is somehow different. How can we explain faith in God by means of psychology or biology if God is not a biological or psychological fact? Similarly, if religion and science are two alternative views of the world, either warring or entirely incompatible, how can one explain the other?

I do not believe we have the luxury of avoiding this scrutiny much longer. The coming nuclear showdown demands that we find scientific and testable answers to a few critical questions about religion and about the world. First, what is it that truly motivates people to be religious? And second, when do religious feelings and choices lead to self-destruction? In any attempt to answer these questions with precision it is vitally important that we take religion seriously and that we avoid oversimplifying what we wish to understand. All too often this is not the case among scientists who try to understand religion.

Why Be Religious?

Research demonstrates that with the possible exception of a few European countries, the number of people worldwide who identify themselves as religious is on the rise. A few well-publicized recent books have attributed the rise in religious identification to ignorance or stupidity, but clearly this is not a very scientific explanation. Far more nuanced have been the analyses of researchers who study broad religious trends by means of concepts like globalization, migration, and postcolonial identity. But anyone who simply listens to people talking about their faith can arrive at a simpler and more empathetic reason for the rise of religious activity: religion, in the many forms it takes around the world, tends to make people feel happier than its absence. The religious life seems to give people a greater purpose, a sense of belonging to a community, perhaps even a confidence in God's love for them and the whole world.

So this is the short answer: religion makes us feel good. In fact, when was the last time you heard someone say something along these lines: "Faith in God makes me miserable, but duty compels me to believe"? There are other occasions for reluctant, perhaps even heroic, duty: volunteering for the army, serving food in a soup kitchen, donating a kidney. Religion is different. And for that matter, when was the last time someone told you she was religious because faith in God helped her understand how the oceans were formed or when the drought will end in Arizona? I suspect that even fundamentalists who look to religion for answers to such questions would drop it in a moment if religion made them miserable. They would deny, of course, that this could ever be the case. So happiness or joy is the gorilla in the room when it comes to religion. In my opinion, religious happiness trumps religious knowledge—religion is not primarily about explaining reality, unless the explanation is itself satisfying. And similarly, people are not religious out of a sense of duty. Religion functions differently from duty: what makes religion compelling is the deep and abiding joy it provides.

But why do so many people feel good about their religion or their faith in God? Is this feeling based on a dispassionate examination of the evidence that led to the conclusion that God does in fact love the world, or that life does have some higher purpose that includes one's own well-being? I suspect that the feeling is so deeply engrained that it is virtually built into our cultural DNA. The positive feeling that attends religion—not the idea of God but the feeling that goes with faith in God—is a product of natural and cultural evolution. Just as maternal love has evolved biologically and culturally, so has the pleasure that religious people derive from their religious feelings. This is why religion has emerged in human history and why it is so persistent in the twenty-first century and for the foreseeable future.

This book will take readers on a long journey, from the biological origins of pleasure to the heights of mystical raptures and to violent self-sacrifice for the glory of God. The stunning tenacity of suicide bombers and the culture that sends them to die will make perfect sense in the light of evolutionary theory. The thread running through the distant stations of this journey is pleasure—first in its simplest biological or adaptive functions, where it contributes to biological survival, then through increasingly complex and culturally constructed forms. The journey will be marked by surprising insights: it turns out that pleasure is not always what we expect it to be. For example, we normally think of pleasure as sensory, emotional, intellectual, or spiritual. I will show that the evolutionary biology of pleasure would have us divide it instead into novelty, mastery (both achievement and self-control), and exploration. These dimensions are relevant not only to humans but to animals as well. For instance, a mouse will quickly learn to solve a maze-puzzle for a piece of cheese (novelty pleasure), but after two or three successful attempts will become bored—the novelty wears off quickly—and begin to explore outside the maze. The mouse's adaptation to novelty leads it to seek a new kind of pleasurable stimulation (exploration) once it has learned the task (mastery). This is why my daughter had to be entertained after three bites when she was fed as an infant. Mastery, for both mice and infants, is exhibited in the capacity to defer the onslaught of boredom and pursue new novelty.

I argue in this book that religion has emerged and prospered as an institution in human history because of the way it controls natural pleasures—sex, eating, and money come to mind—and allows them to evolve in response to increasingly complicated social situations. The many religious traditions around the world implement various techniques for overcoming human enslavement to novelty pleasure, which tends to be individualistic and "selfish," and turning this pleasure into more elevated mastery enjoyments, which are social. Thus, sex, eating, and money become marriage, sustenance, and charity. Religion's techniques for achieving such transformations can often look like simple morality: "Don't have sex unless you have to and always eat moderately" is an injunction common to many sacred scriptures around the world. Even to a modern writer like C. S. Lewis, the joys of a durable marriage far outweighed the pleasures of a new sexual encounter. In this book, we learn to look at religion's techniques not as moral rules but as hedonic manipulations.

"Hedonics" refers to the science of pleasure; the Greek Hedonists did not simply chase pleasure—they also refined it and theorized about it, as do many scientists today in the fields of psychology and economics. Through hedonic manipulation, religion achieves the central task of turning selfish individuals into willing, even altruistic, members of a group. Religion does this best because its sophisticated hedonic techniques make us happy. Self-control is the secret to deep and lasting happiness! The Dalai Lama, to give one recent example, argues in his bestselling book *The Art of Happiness* that the purpose of life is to gain happiness and that we do so by controlling our need for mere pleasure.

Once we understand how religion makes us happy, we can begin to explain many of the things that religions around the world have done over the millennia. Some of these religious events and accomplishments, such as the Sabbath meal in Judaism, Persian mystical poetry, or carnivals in Rio or Venice, are wonderful. Others are also grounded in the hedonic techniques of religion but are not impressive in the same way—such as, at one extreme, the man or woman who finds joy in obedience to a spiritual master that leads to a fiery self-annihilation in the name of God.

As the book progresses, we look at the frequently surprising history of how complex joys have emerged from simple pleasure. For example, the early Christian was far more likely to take his food without spices than either the Jew or the Greek. Four centuries earlier, the first-generation Buddhist felt the same way as the Christian about his dining, while the Hindu did not. Focusing on pleasures around the table reveals hidden patterns in the religions of the world: some religious groups (Jews, Hindus) encourage their members to embrace the pleasures of life, while differently structured groups curtail such pleasures for the sake of something they regard as loftier—a more abstract joy that requires the giving up of spices, sweets, and sauces. What accounts for these sometimes surprising differences is not the nature of the sacred doctrine or object of faith (God, Dhamma) but the social organization. We discover, for example, that the more voluntary an adherent's membership in a religious group is—whether by converting, becoming a monk, or joining a world church—the greater the group's insistence that members give up simple (novelty) pleasure in favor of higher pleasures such as mastery. Readers will come to understand how this explains, in a roundabout way, why religious suicides are more likely to emerge from Islam than Judaism, and why spiritual devotion is more dangerous than blind observance of religious law.

We shall encounter other surprising ideas. Mystics—men and women who seek direct communion with God—have at times attained profound states of joy. Some of these individuals, like Teresa of Ávila, Saint John of the Cross, or Jalal al-Din Rumi, have gained worldwide fame for their articulate descriptions of divine raptures. The joy, they tell us, is divine grace, a gift of God's overwhelming love. It is difficult to argue that religion is dangerous when it speaks of love, and in fact most critics of religion regard its potential for violence *as a separate matter* from mystical raptures and divine love. I do not.

To understand what makes religion truly dangerous in a nuclear scenario, we need to realize that religion at its very best—at the place where grace and love meet patience and humility—is, astonishingly, where the seeds of self-annihilation may sprout. The feelings of love that great mystics report as emanating from God at the moment of supreme spiritual achievement are based on an error that is unique to pleasure. I call it the "Prozac effect" throughout the book. Prozac and other pharmaceutical antidepressants—or, for that matter, a double espresso—can have profound effects on mood but remain beneath the level of awareness, "in the brain." It is the world that seems forgiving and wonderful, not the serotonin in the synapses or the caffeine in the bloodstream. Such a misperception also accounts for the ways in which pleasure can motivate us while eluding our conscious awareness. The Prozac effect on the great mystics was to generate not only the certainty of God's existence but also a powerful feeling of love that seemed to radiate from God. Such love is wonderful to experience and far surpasses anything as mundane as the social contract or the notion that one should accommodate others only because of a sense of duty or mutual obligation. Such love can also be dangerous.

This divine love attained by a charismatic spiritual leader has often attracted groups of followers who are mesmerized by the personality of the spiritual master. Love or charisma becomes the organizing principle for the master-disciple group. Ironically, it is this love—a form of pleasure, in fact—that can lead members of such groups to self-annihilate in acts of revolt or martyrdom. And what is the reason for the violence? A society based on laws and obligations, even if these happen to be religious (Shari'a or Halakhah), cannot live up to the highest spiritual standards. Only love does that. Devotion to God trumps the Golden Rule.

Readers may wonder whether I really think that pleasure is the most important or most interesting feature of the religious life. The answer is no, I do not. But pleasure is both essential to religion and subject to scientific study. The ultimate purpose of this project is to weigh in a scientific manner the likely responses of a religious actor in a crisis that involves the potential for high levels of violence. We need to know, for instance, whether deterrence will be effective: Is the religious actor rational? Is he interested in maximizing utility (where this does not refer to an apocalypse but to prosperity in this world)? The study of hedonics from the psychological, economic, and political science perspectives can now be extended to religion. By tracing the uses and elaborations of pleasure in religious contexts, we discover that apocalyptic beliefs are not nearly as dangerous as we might think; that in some situations the Iranian mullahs, with their Islamic laws, pose far less danger than a lovingly disposed spiritual master, whether Muslim, Jewish, or Hindu; and that the way a religious group clusters is more important than what its members believe or feel about another group.

To return to the metaphor of the two drivers playing chicken, the science of hedonics applied to a religious driver can tell us whether he will swerve or seek the ultimate collision. Making that determination is the practical and urgent goal of this book. But equally interesting, I believe, is the scientific—biological and psychological—explanation offered here for something essential in the religious life of humans: religious pleasure.

CHAPTER 1

Religious Self-Destructiveness and Nuclear Deterrence

ON AUGUST 9, 2001, AT around two o'clock in the afternoon, a little after peak lunch hour, the Sbarro Pizzeria on King George Street in Jerusalem was full of customers. Families and young people crammed the counter and occupied virtually all the seats. Just then a man walked into the restaurant, Izz al-Din Shuheil al-Masri, the twenty-two-year-old son of a restaurant owner from the West Bank. Under his loose shirt, al-Masri was wearing a belt packed with nails, nuts, and bolts and an explosive device featuring a simple detonation mechanism. Well inside the pizzeria, he set off the belt, killing 15 people and wounding 130. The dead ranged from two-year-old Hemda Schijveschuurda to sixty-year-old Giora Balash.[1]

Six weeks later—and exactly two weeks after the 9/11 terrorist attacks in the United States—the *New York Times* reported that the Palestinian students of An-Najah University in Nablus, on the West Bank, celebrated the Sbarro massacre during a campus event by erecting a model of the restaurant, with broken furniture, shattered body parts, and mock blood splattered on the walls.[2]

Sacred Suicide

THE ATTACK ON THE Sbarro diners, like many other suicide bombings, shocks and dismays us beyond most other crimes. This reaction is due not just to the level of violence the bomber inflicted; we have also been horrified by the suicide bombers who fail to kill anyone but themselves, or who injure only a few checkpoint soldiers. More striking even than the level of violence is the nature of the killer—the man who moves among his victims, old people and infants, brushes against them and perhaps even smiles, before blowing up himself and everyone around him. The suicide bomber crosses a line that we draw to mark off humanity—even human conflict—from a level of evil in which the perpetrator has no empathy for the life of his victims because he detests life itself. The exhibit at An-Najah testifies that this evil can actually be celebrated, confirming our darkest suspicions.

Perhaps as a result of such horrific acts, some have looked to explain them, if not justify them, as the fruition of personal despair and extreme anguish: If you grow up in a refugee camp, in squalor, with no work, land, or future to speak of, death, gloriously violent death, can seem like an appealing act of last resort.[3] All the experts now know, however, that suicide bombings have little to do with personal despair. Al-Masri grew up in a comfortable household; Hamadi Jaradat, the twenty-seven-year-old female bomber of Maxim Restaurant in Haifa on October 3, 2003, was a lawyer; and most famously, Muhammad Atta, the leader of the 9/11 suicide terrorist group, was a well-educated, smart, and successful urban landscape professional in Germany when he took on his mission.[4]

The natural liberal impulse to explain crime in terms of socio-economic factors collapses here, and so does its assurance of rational remedies. We are left with a sense of cultural bewilderment, as though we have encountered a new and unknown civilization that we must come to understand in order to stop these killings. But how do we explain a cultural phenomenon in which the following songs are composed by the relatives, and even the mothers, of these killers?

The martyr gives us stones from his blood
From his red blood the rose becomes red
His mother trills for him in joy
He has given his blood to the nation.[5]

Or consider the following recitation by a young woman on Palestinian television:

Good-bye, my mother, father, brother,
Most beloved among humans.
Do not grieve for me if you see me return
Carried on shoulders.
Do not cry for me if I die a shahid [martyr],
For my joy at meeting
My God is the most beautiful of promises.

There is not one word here about the tragedy of self-sacrifice or the painful and reluctant obedience to the higher calling of duty. Instead, there is a mixture of joy and religious devotion that turns a gruesome act of murder and self-destruction into a pseudo-mystical religious celebration.

There is nothing unusual about a society that glorifies and even celebrates those who sacrifice their lives for their country, community, or religion. Most nations set aside a holiday to mark such sacrifice. We have our own Memorial Day, for example, with flags, wreaths, motorcycle parades, and pilgrimages to the burial places of our fallen military heroes. During these holidays, we grieve for the recently killed, but we also celebrate and remember the earlier generations of soldiers who gave their lives for us. The fact that suicide bombers are murderers in the eyes of the world does not diminish their prestige within the society that feels served by their suicide—their sacrifice. Morality aside, all societies celebrate their heroes. It is only natural.

In fact, even some animal societies display self-sacrificing behavior for the sake of the group. The Memorial Day edition of the *Washington Post* on May 28, 2007—the same issue that covered the ceremonies in Arlington National Cemetery—also ran a short piece in the "Science Notebook" section about army ants, which lay their bodies down over potholes so that the rest of the force can proceed across the hazard. Bees also behave in a self-sacrificing, "altruistic" way by dying for the hive.[6] Many species of birds have individuals that scan the horizon and send out alarm calls, at great risk to themselves, when they spy a predator.[7]

The reader will protest that animal altruism is a matter of genetic programming; the individual exercises no true choice in the matter and therefore does not genuinely—and certainly does not morally—practice self-sacrifice. This rings true and seems to fortify a very sharp boundary between fully human and animal behavior. We have "duty" and "honor" and "love of country" and "faith," while animals have stimulus, response, impulse, instinct, and mere behavior. So why bring up the subject of animal altruism? Oddly, that same Memorial Day edition of the *Post* ran a front-page article by Shankar Vedantam, the staff writer who specializes in finding thrilling scientific stories. His piece was titled "If It Feels Good to Be Good, It Might Be Only Natural." The article reviewed the experimental work of a number of brain scientists, including Jorge Moll and Jordan Graffman, who discovered that generous actions "activated a primitive part of the brain that usually lights up in response to food and sex." Put simply, behavior that favors others or the group as a whole—in a word, altruism—is hardwired into our brain. Moreover, the altruistic behavior must have evolved successfully because it was backed up by that ultimate action guide, namely, pleasure. We do what feels good—evolution has seen to it that it benefits the group.[8]

Evolutionary biologists, including Darwin, have long argued that morality evolved naturally among the advanced species. Recently published books in evolutionary ethics have been generating heated debates among biologists, philosophers, and theologians. The biological argument makes human morality, including altruism and self-sacrifice, naturally entrenched, which does not bode well for eliminating extreme displays of group-favoring behavior like suicide terror. At this time it remains doubtful that philosophers, economists, and sociologists who champion free will and "human dignity" will find the empirical research of biologists persuasive.[9] Nor will any of them attribute suicide terror to biological impulses. Their trump counterargument runs something like this: Since humans are free to ignore "good feelings" (pleasure) when it comes to gluttony, sloth, incest, and many other social sins, it certainly would be far easier for them to choose to ignore the pleasures of virtue (self-sacrifice). The decision to do good can thus never be based on the feeling that accompanies the act, but only on the rationale for doing the right thing. This is a powerful argument against biological determinism and a serious obstacle for writers, such as myself, who wish to argue that biology can explain—at least in principle—all forms of human behavior.

The true basis of acts of ultimate self-sacrifice must remain open for the time being. What is clear is that suicide terrorists' commission of a crime is hardly the only way in which they differ from military casualties. There are three basic differences:

1. Unlike the soldier, the suicide terrorist succeeds only when he dies. Death is the necessary means to his end.

2. The rhetoric of the suicide terrorist is often saturated with strong religious imagery.

3. The personal statements of suicide terrorists go beyond reluctant, duty-bound calls to service and describe a joyful pursuit of supreme felicity.

In a recent report by Anne Marie Oliver and Paul F. Steinberg, a Palestinian named 'Uthman, who was a failed suicide bomber, describes a mission to hijack and blow up a bus in Jerusalem. Before the attack, a handler of 'Uthman and two other young volunteers had arrived to announce the nature and time of their mission and made the following promise: "You're now going to be with the Messenger of Allah and 'Umar and Abu-Bakr the Truthsayer and the martyrs." The mission virtually precluded any possibility of survival, but this apparently failed to scare 'Uthman, who told his interviewer: "You know, we were to be gathered with the martyrs, so we were overjoyed."[10] Earlier in the interview, the young Palestinian had used an extraordinary metaphor to describe his feelings before the suicide mission:

> On the night before our operation, we had that feeling when you first get married, on the night of the wedding, so excited.

> . . . That feeling comes from the fact that soon you will see the Gardens of Heaven; you will see the Prophet and your friends who died before. We were so excited, and we were full of happiness.[11]

'Uthman survived his failed attack on the bus, and his brave words, spoken in prison, may reflect the bravado of a man relieved to be breathing. Still, the three markers are there in his narrative: certain death, religious imagery, joy. In fact, his testimony gives credence to the speculations—some say rantings—of the British novelist Martin Amis, who claimed that suicide bombers act out of sexual frustration in a culture that does not allow them free sexual expression.[12] How else would we explain the sublime reward of the seventy-two heavenly virgins who await the shahid (martyr)? Even the mother of one terrorist (Hamza Samudi) told an interviewer: "Hamza wanted to bring me into paradise. . . . He would take beautiful women, the most beautiful women in paradise; he lived and died like a hero, a blessed hero."[13] To those who point out that young women and even mothers have strapped on the explosive belt, Amis could argue that the lawyer from Jenin who blew up Maxim was a twenty-seven-year-old unmarried woman—that is, someone very likely to remain a spinster. Moreover, mothers have volunteered to die when suspicions were cast on their marital fidelity.

The vast majority of experts on terrorism—those who are paid to analyze the phenomenon and come up with a plausible profile of the suicide terrorist—reject any one simple or reductive factor. They have cited personal reasons, such as revenge or depression, and cultural reasons such as honor as a family value, the economics and psychology of unemployment, politics, strategic warfare, and religious beliefs. Whatever the guiding motives of the suicide terrorist—and often they are mixed—an efficient and sophisticated machinery is in place to identify potential volunteers, deepen their commitment to kill and die, guide them to the final act, train them to operate the belt, teach them to ignore any counterproductive feelings such as fear or empathy, transport them to the target, and so forth. An increasing number of researchers such as Shari Goldstein in Israel and Hector Qirko and Scott Atran in the United States have described the elaborate psychological work that goes into the making of a human bomb. Although religion is extremely important—how could it not be?—simply being a devout Muslim is never enough.[14]

In fact, suicide terrorism is hardly unique to the Islamic world. Next to Hebron, one of the major cities in the Palestinian West Bank, is the tomb of a Jewish man that reads:

> Here lies the saint, Dr. Baruch Kappel Goldstein, blessed be the memory of the righteous and holy man, may the Lord avenge his blood, who devoted his soul to the Jews, Jewish religion and Jewish land. His hands are innocent and his heart is pure. He was killed as a martyr of God on the 14th of Adar, Purim, in the year 5754 (1994).

The inscription reads as if Goldstein was a victim of terror. In fact, on the day stated, Baruch Goldstein, an American-born Israeli physician, armed with an AK-47 assault weapon and several clips, entered the Ibrahimi Mosque during prayer time and opened fire on the worshippers. He kept firing for about ten minutes, until he ran out of ammunition. The toll was 29 dead and 150 wounded men and boys. Goldstein's attack was suicidal inasmuch as he did not retreat while firing. He simply finished shooting and remained in the mosque. The survivors proceeded to beat him to death.[15]

There have been other religiously inspired attacks by Jews against Arabs, but not suicidal ones. In contrast, self-destruction is at the very heart of the terrorism perpetrated in South Asia by the Sri Lanka Tamils, formally known as the Liberation Tigers of Tamil Ealam (LTTE). Their most spectacular success took place in India on May 21, 1991. That was the day Rajiv Gandhi was assassinated by a female suicide terrorist, named Gayatri or Dhanu Rajaratnam, during an electoral campaign stop. Gandhi was slowly making his way to the speaking stage, surrounded by well-wishers who were placing traditional silk scarves around his neck. A bespectacled young woman who had not been cleared by security approached bearing a sandalwood garland in her hand. She bent down to touch Gandhi's feet in a gesture of respect, then set off the improvised explosive device (IED) she was wearing. Eighteen people were killed and thirty-three injured.

Gandhi's crime had been sending Indian troops to Sri Lanka to broker a truce between the Sinhalese army and the LTTE. Years later, the LTTE assumed responsibility for the assassination and apologized for it, using such terms as "great tragedy" to describe the death of Gandhi.[16] But in Sri Lanka itself, the LTTE had been waging a war of terror since the 1970s toward the goal of national independence for the Tamil majority (South Indians) in the northern regions of Sri Lanka—an island-nation where the great majority of the population is Sinhalese. The LTTE had adopted methods that made Hamas and Islamic Jihad pale in comparison. In addition to assassinations and attacks on civilians (children and infants not excluded), the Tamils have engaged in widespread atrocities such as ethnic cleansing, execution of prisoners, conscription of child-soldiers, and extortion. But suicide terrorist attacks by so-called Black Tigers, either male or female volunteers, has been their trademark.

Peter Schalk of Uppsala University in Sweden has analyzed the ideas driving the Tamil independence movement, especially as formulated by Velupillai Pirapakaran, their leading figure.[17] Schalk finds a deep mixture of politics, religion, and even mysticism among the Tamils. Surprisingly, the goal of independence (*cutantiram*) has been rendered sacred in a pseudo-Jewish manner: it promises a Zion-like land to a select people. Indeed, the early leader of the movement, S. J. V. Celvanayakam (1898–1977), was hailed as a Moses figure who did not live long enough to cross over into the Holy Land. These strangely Hebraic notions reached the Tamils indirectly via the Christianity to which they had converted. Their ideas represent a strange mixture of South Indian Hinduism and biblical Christianity. For example, self-sacrifice in service of the ultimate political goal is considered martyrdom. Such an act requires a uniquely devotional attitude called *arppanippu* in which the sacrifice of one's life for a holy cause is welcomed with open arms. The moral implications of the violence play virtually no role in assessing the sanctity of the act or the spiritual qualities of the terrorist. However, a small number of nonviolent martyrs (including Tilpan and Pupati) who fasted to death in a manner that is usually described as Gandhian are also celebrated among the Tamils of Sri Lanka.

Religion and the Bomb

Many experts on suicide terrorism regard it as a "top-down" phenomenon. In this view, since neither personal circumstances nor individual psychology would ever naturally lead a person to think of strapping on an explosive belt and blowing himself or herself up among innocent civilians, suicide terror is not suicide but rather a strategic use of human missiles—one that requires a guidance system, so to speak. Some terrorist organization must decide that launching people as walking bombs is a terrific way to fight a war, and there is nothing spontaneous or emotional, let alone suicidal, about this organization.[18] Deterrence or prevention is thus not a matter of psychology or sociology in the way that dealing with teenage suicide, even the murderous suicide at Virginia Tech in April 2007, might be. The cold-blooded and systematic willingness to destroy one's own people, even in large numbers, for the sake of strategic ends becomes an enormously complicated threat. The best case in point here—and the most frightening prospect for the use of violent self-destruction—is Iran. That country has launched an ambitious plan for developing nuclear arms and, despite international pressure, will probably attain its goal in the next decade. Whether Iran will use its nuclear weapons, specifically against Israel, is difficult to say. A likely consequence of Iran attacking Israel, however, for whatever political or strategic reasons, would be a devastating counterattack.

Up to now, cold reason, common sense, and the simple will to survive have helped to maintain the balance struck between nuclear powers, such as the United States and the Soviet Union during the cold war. MAD, or mutually assured destruction, was a macabre but effective hedge against aggression. You do not attack if you are certain of your own destruction. But would such a simple calculus work when one side, flush with a profound religious-messianic ideology, rejects the line between self-destruction and sacred sacrifice? Is Iran the place where such a strong religious impulse prevails?[19]

It is hard to say how Iran would behave if cornered in a nuclear showdown with another country or with the West. There are a few indications, however, from its past. During the Iran-Iraq War of 1980–88—a war that was largely ignored outside the Middle East—the Iranians supplemented their regular army and the Revolutionary Guard with a huge cadre of volunteers called *Basij* ("mobilized"). Millions of civilians, including tens of thousands of teenage boys, driven by the spirit of the Islamic revolution and inspired by the

Ayatollah Khomeini, gladly offered up their lives in sacrifice for the nation. Vast numbers of these volunteers, poorly trained and badly armed, died in ill-conceived assaults on Iraqi positions that were fortified with artillery and armor. Many of these children were used to clear mines or wade through swamps in the Basra region. One twelve-year-old Basij boy, Hossein Fahmideh, became the symbol of self-sacrifice for blowing himself up underneath an Iraqi tank.

The ideology of Basij, which is both Islamic and nationalistic in its view of self-sacrifice, has become the guiding spirit of Iranian rhetoric. There are nine million Basij Iranians today, including President Mahmoud Ahmadinejad, who has defined Basij as "a willingness to die for one's homeland." In 1995, during the annual Basij week, the Iranian supreme leader, Ali Khamenei, stated: "The Basij is the essence of the regime's functioning." In another speech, he was more specific: "The stronger the Basij become, the more secure our country will be in the future. . . . The reason for the success of the Basij is its members' faith and trust in God."

Other countries also praise those who make the ultimate sacrifice, and even Americans play up the role of faith in God as a national rallying cry. Indeed, none of the statements just quoted expresses a positive desire for self-annihilation. But if the fairly limited aggression of Iraqi leader Saddam Hussein could trigger in Iran a religious-nationalistic fervor capable of producing an organization that efficiently drafted vast numbers to a truly senseless death, what would a nuclear standoff do? If victory in war, as the military theorist Carl von Clausewitz defined it, depends on the ability to sustain greater levels of pain, could the Iranians—headed by a Basij member—be counted on to back down to avoid extreme destruction to their country in a nuclear exchange? Or, to put this more parochially, would you wish to play "chicken" with a Basij?

Unfortunately, a day will come when such a game begins. The Iranian government may decide that American naval activities in the Persian Gulf or air bases on the Arabian Peninsula represent a national threat. Or the Iranians may decide to end the very existence of Israel. In fact, on the eighteenth anniversary of Ayatollah Khomeini's death, President Ahmadinejad announced that the countdown to the destruction of Israel had already begun. The question of the willingness to self-destruct or to sustain devastating pain will come up in any realistic assessment of ways to deter the Iranians from ratcheting up a nuclear standoff to the point of no return.

To deter the Iranians effectively, the role of religion will also have to be assessed along with all the other factors involved. There is something about religious attitudes and feelings—and this is not unique to any specific religion—that makes humans behave in odd and apparently counterproductive ways. What is most intriguing here is not the role of religion in propelling humans to act aggressively and violently, a subject that has been studied extensively in recent years. More puzzling is the function of religion in human willingness to self-destruct, sometimes even joyfully. It seems that religion is unique in human culture as a positively conceived way of undermining one's own existence. But only by understanding such self-destructive aspects of religious belief can the challenge of deterrence be met.

Religious Self-Destructiveness

Religiously inspired suicide terrorists and rogue regimes come under the much broader heading of self-destructive religious behavior by individuals and communities. You need not act aggressively against others in order to undermine your own well-being and prosperity. Throughout history members of religious groups have injured their own health by means of fasts, vigils, and other austerities. Many Christian Scientists today risk death by forgoing medical treatment. Families have broken up when members left to join monasteries, follow cults, or adopt gurus. Men and women have declared themselves saviors, prophets, and messiahs and led their followers to financial and even physical ruin. Space permits only two detailed examples, both of them perhaps surprising ones. In fact, these two cases demonstrate that the problem with religion is hardly limited to aggression. Just as troubling is the self-destructiveness of the religious pacifist—the self-aggressor.

The first case is the disturbing matter of Orthodox Satmar Jews attending Iranian president Ahmadinejad's recent conference of Holocaust deniers. The second is Mohandas K. Gandhi's form of self-destructive behavior, which no one would ever consider aggressive. The thesis here is simple: what is significant about religious self-destruction is not the moral component (how many people you take with you as you self-destruct) but the psychological one. What is the frame of mind of those brought to self-destruction by their religious faith? And what kind of general lesson can we extend to the religious terrorist who must be stopped before he sets off the explosive belt?

On December 11, 2006, Iranian president Ahmadinejad hosted a conference with the stated purpose of exploding the myth of the Holocaust. Among the usual Holocaust deniers, neo-Nazis, and Ku Klux Klan (KKK) members (such as David Duke) attending the conference was a small number of men from Rockland County, New York, and Manchester, England, led by Rabbi Yisroel Dovid Weiss. The most striking image from the Iranian conference of anti-Semites was that of Ahmadinejad hugging an Orthodox Jew, Moishe Arye Friedman. The Jewish men represented a small, ultra-orthodox splinter group of Hasidism that calls itself Neturei Karta (Guardians of the City). In Israel they live in a Jerusalem neighborhood called Meah She'arim (One Hundred Gates) and are led by Yisrael Hirsch. Hirsch himself is no ordinary Jew: according to the *Jerusalem Post* (December 19, 2006), the Israeli army seized official Palestinian documents in 2002 showing that Hirsch had been on the payroll of Yasser Arafat's Palestine Liberation Organization (PLO) for years.

What would drive Orthodox Jews, who live according to the laws and traditions of Judaism, to support the most virulent forms of anti-Semitism today? Would they also have helped Hitler in his day, or is there something special about the Iranian version of anti-Semitism?

It is important to mention that on the issue of Holocaust denial (or tacit support given to deniers), Neturei Karta represented a tiny percentage not only of Orthodox Jews but also of the Hasidic movement (Satmar) out of which they emerged. In fact, the leaders of Satmar Hasidism excoriated Neturei Karta in the strongest terms possible: "Those who went to Iran trampled on the memory of their ancestors and people. They embraced the disciples and followers of their murderers." The official publication in Jerusalem of the ultra-orthodox (Hareidi) community described those Jews who attended as a "tiny group of insane Jews who are liable to incite hatred against hareidi Jews."[20]

The attacks on Neturei Karta for their participation in Ahmadinejad's conference cannot completely mask a puzzling fact, namely, that even mainstream Hasidic Jews—both Satmar and Lubavitch—are fiercely opposed to Zionism. The political and national program to establish and secure a homeland for the Jews in Israel has been anathema to anti-Zionist Hasidic Judaism. Hasidism itself, founded by Rabbi Yisrael ben Eliezer (known as Baal Shem Tov, 1698–1760), was a devout and often ecstatic religious movement that emerged in eastern Europe in the eighteenth century in response to what it regarded as the overly dry and legalistic qualities of traditional Judaism. Later, in the nineteenth and twentieth centuries, at a time when the majority of Jews in Europe struggled against the physical threat of annihilation and looked to reestablish a homeland in Israel, messianic Hasidic communities went against the grain. Led by such figures as Rabbi Chaim Elazar Shapira from Monkatash (Hungary), these Jews opposed any effort, especially by Orthodox Jews, to settle the ancient homeland.[21] Shapira called such a place "Zionistan." A follower of Rabbi Shapira, Rabbi Yoel Moshe Teitelbaum, who was the founder of the Satmar Hasidic community, later took over the anti-Zionist campaign. The ultra-orthodox anti-Zionist groups responded to the destruction of European Judaism by moving to New York and Brooklyn. Chaim Potok's *The Chosen* explores the clash there between those Jews who supported the founding of the state of Israel after the war and those Hasidic Jews who opposed it. " 'The land of Abraham, Isaac, and Jacob should be built by Jewish goyim, by contaminated men,' Reb Saunders shouted again. 'Never! Not while I live!' " (187).

The anti-Zionists offered a simple argument: The long-standing exile of the Jews is a punishment meted out by God, and as such it must not only be tolerated but actually loved! The diaspora will gather in the Holy Land only with the coming of the Messiah; any effort to rebuild Israel in the meantime is a "hastening of the end." This argument is based on a familiar Talmudic quote and discussion of a verse from the biblical Song of Songs: "I abjure you, O maidens of Jerusalem, by the gazelles and by the hinds of the fields: do not wake or rouse love until it pleases to come of itself." Anything short of messianic redemption, not the false ones of pretenders like Sabbetai Tzvi, constitutes a sin. This is not to say that the Satmar wish to see the destruction of Jews in exile. On the contrary, they look forward to the complete and final redemption of the Jewish people. But because that depends on God alone, and because for the time being God wills the exile of the Jews, any positive act of national self-help is forbidden.

Here, then, is a clear case of religious self-destructiveness—or, at the very least, counterproductive and maladaptive beliefs. It may seem grossly unfair to compare the position taken by anti-Zionist Jews with the aggression of suicide terrorists or the Basij of Iran, but underlying all of their beliefs is a unique religious value: the ultimate joy anticipated in the future depends on and justifies the risks and destructions of the present time.

If the Hasidic Jews make for a surprising case, consider Mahatma Gandhi. Mohandas (Mahatma) Gandhi is almost universally revered as one of the great leaders of the twentieth century. Among many outstanding political figures, he certainly enjoys the reputation of the saintly one—the man who defeated power with love, who applied his religious philosophy of nonviolence to end the rule of the British Empire in India. Nowhere else in the turbulent twentieth century was so much achieved with so few deaths. However, we may justifiably question the true effectiveness—even the rationality—of Gandhi's centerpiece idea: *satyagraha* (force of truth or nonviolent resistance) as a practical method for combating evil. Imagine that the occupiers of India had not been Eton- and Oxford-trained civil servants but Nazis—would Gandhi's methods have made sense? Would he have clung to them in the face of their predictable failure?

We need not speculate to answer these questions with a no and a yes, respectively. In 1938, just after the Kristallnacht attacks on Jews throughout the German Reich and shortly before the war and the systematic eradication of Europe's Jewry began, Gandhi published an open letter on the subject of the Jews under Nazi rule. He fully recognized and lamented the persecution of the Jews in

Germany. If ever a violent resistance was justified, he conceded, this was the case. But violence was never justifiable, he asserted, not under any circumstances. In fact, not only did Gandhi reject violence, but he even condemned Jewish emigration to a safe homeland. As a matter of principle, as an example of true satyagraha, only voluntary submission to mistreatment—even to murderous violence—would bring the Jews "inner strength and joy." He added that if the Jews could train their minds properly (like Gandhi's own followers), even a complete annihilation of Germany's Jewry "could be turned into a day of thanksgiving and joy."[22]

Gandhi's letter provoked an angry response from the dovish theologian Martin Buber, who argued in a letter of February 24, 1939, that the great Indian leader completely failed to understand the gravity of the situation facing the Jews of Europe.[23] A reader today might be inclined to feel more cynical than Buber and assume that Gandhi simply did not care for Jews. Not that he was anti-Semitic—simply distant and unconcerned. This would be factually incorrect. Furthermore, it would seriously underestimate the unbending tenacity of Gandhi's beliefs and the depths of his religiously inspired self-destructiveness. The proof is in the manner in which he addressed his fellow Hindus in similar circumstances.

In the summer of 1947, having partitioned the Indian subcontinent along communal-religious lines into India and Pakistan, the British packed up and departed. What ensued was a massive and bloody ethnic cleansing by both the Hindu and Muslim communities (with the Sikhs siding principally with the former). Driven by the hideous violence, millions of refugees streamed across the new border in both directions: Muslims escaping to Pakistan and Hindus avoiding death by moving east to India. Gandhi, who had opposed the partition all along, was deeply disturbed by the forced migration of both communities. To the Hindus who were leaving ancient Indian cities now in Pakistan—cities such as Lahore—he addressed the following words: "If you think Lahore is dead or dying, do not run away from it, but die with what you think is dying in Lahore." A few months earlier, Gandhi also made the following statement: "I would tell the Hindu to face death cheerfully if the Muslims are out to kill him. . . . You may turn round and ask whether all Hindus and all Sikhs should die. Yes, I would say. Such martyrdom will not be in vain."[24]

Oddly, Gandhi was telling the Jews in Germany and the Hindus in Pakistan to die because he believed in their potential moral superiority over the people who were out to kill them. In Gandhi's view, glad acceptance of voluntary destruction could thus act as a kind of lesson to the killers—a form of spiritual education about the nature of true joy. Gandhi demonstrated his own lifelong commitment to this form of martyrdom by repeatedly threatening to kill himself through fasting. The slow-burn suicide of fasting was a way to destroy the body and thereby inflict grief, shame, or guilt on an opponent who might be susceptible to what Erik Erikson had correctly termed "blackmail."[25] The targets of this anticipated suicide-by-hunger could have been the British colonizers of India, but far more frequently they were people who were closer to Gandhi, who loved and respected him. A famous example of such a fast was the labor strike at the Ahmedabad mill, whose owner—Sheth Ambalal—was a good friend of Gandhi's. Gandhi announced that he would not eat until the workers' demands were met—and so they were. In another, perhaps more troubling example, Gandhi had uncovered some minor moral misbehavior by two young men and a young woman, all of them about twenty years old, who were his disciples at the Phoenix farm in South Africa. To punish the young people, Gandhi decided to fast: "I felt that the only way the guilty parties could be made to realize my distress and depth of their own fall would be for me to do some penance. So I imposed upon myself a fast for seven days and a vow to have only one meal a day for a period of four months and a half." The girl promptly shaved her head and joined him in the fast.[26]

Would it be fair to condemn Gandhi's acts of self-aggression as counterproductive displays if his goal was the education of others? The Pulitzer Prize–winning biographer of Gandhi, the psychologist Erik Erikson, thought so: "Here self-suffering could harbor the despotism of a cruel (if "cruelly kind") father who, by his self-suffering, hurts ever so much more vindictively, and ever so much more unfathomably than an outright angry one; whereupon the children feel punished if not 'crushed'—but by no means persuaded."[27] Gandhi was probably deeply influenced by his mother's many fasts—and probably also by her passive-aggressive interactions with the family. It is probably unfair to regard Gandhi as cruel, but his understanding of proper religious motivation acted as a double-edged sword: it was both healing and hurting. As a well-considered moral theology, Gandhi's form of "education" could never be extended to the real-world sufferings of individuals or communities. It was far too unbending in its internal logic of self-annihilation.

Readers may object to this juxtaposition of Gandhi's method with those of suicide terrorists. But there is no need to flatten the terrain completely in order to see the similarity between Gandhi, Satmar Jews, Iranian Basij, Tamil Tigers, and Palestinian human bombs. Clearly, Gandhi never killed a soul, and just as clearly the Sbarro killer was not trying to strengthen anyone's moral fiber. But in their distinct ways, all of these are primarily religious actors. All *choose* their own personal disintegration, or that of their community, in order to attain something far grander than pragmatic solutions to practical problems. They seek nothing short of a beatific vision—a religious ideal—that offers up permanent happiness in the future. It could be anything from an idealized land to *moksha* (spiritual liberation). To put the matter technically, all of them are hedonic gamblers of the highest magnitude—the all-or-nothing variety.

It is important to emphasize that although all forms of religion aim at various states of satisfaction, not all forms of religion engage in the same totalistic wager described here. For example, the Passover meal combines culinary pleasure with family togetherness to celebrate spiritual and physical freedom. The Catholic High Mass is visually and musically enjoyable. So is prayer in general. There is no hedonic trade-off in many religious practices, no voluntary suffering for future bliss. We can even understand many types of ascetic behavior without metaphysically wagering the entire farm, so to speak. Pilgrims who undergo physical hardships (walking barefoot, sleeping on hard surfaces, enduring heat or cold) form a special bond as a religious community of co-travelers during the pilgrimage. The benefits, which are emotionally satisfying, gradually emerge as the costs (hardships) progress. In contrast, the thinking of the all-or-nothing religious gambler can seem odd, even bizarre. It insists on a *total* joy that depends on *total* sacrifice. And even this belief that the two sides of the quid pro quo are necessarily connected, despite the delay, is a source of a particular kind of pleasure. To put this in somewhat more subtle terms: such a belief (that joy hinges in a delayed fashion on suffering) removes the unbearable discomfort inflicted by a sharp cognitive dissonance, namely, the thought that the present state of affairs, which may be quite bad, means nothing whatsoever.[28]

Three basic questions must be cleared up here before we proceed: (1) Why is cognitive dissonance so uncomfortable that it must be resolved at all cost? (2) What makes meaningless suffering dissonant? (3) Is pleasure (including happiness and joy) really such an important factor in the religious life of believers? Only by clarifying the basics can we construct a proper way of confronting the religious terrorist in his pursuit of joyful annihilation.

Responding to Religious Self-Destructiveness

IMAGINE YOU ARE THE factory owner who has to negotiate with labor union representatives and with Gandhi. You tell Gandhi that by threatening to kill himself slowly he is being unfair. You know how revered he is in the country, and you also realize that, if his hunger and death are seen as your fault, you will be ruined. This is a reasonable, "rational" assumption, and it may oblige you to go further than you would like in favor of the workers in order to save your business. Clearly, too, Gandhi is a savvy negotiator; he knows his own hand and is ruthless about playing it. But there is a stark difference between your position and Gandhi's: your entire strategy is based on economic prudence, while Gandhi's religious method and rhetoric seem designed to achieve complete "moral" victory. Naturally, your offer fails to satisfy him. Would it be enough to show him your account books? If his take-no-prisoners approach appears to be sensible when you really are being unfair to your workers, does that mean that his approach is always sensible? Is it rational in the sense of being subject to reasoned self-scrutiny and a willingness to settle for something short of perfect justice? For that matter, are any religious actors who engage in totalistic gambles (complete self-destruction in exchange for complete satisfaction) ever rational?

The subject of rational choice theory—figuring out the manner in which people calculate their choices—is an enormous field within economics, political science, psychology, and even religion.[29] The classical approach has been to argue that "rationality" implies acting in a way that maximizes one's self-interest according to certain rules of internal consistency. For example, if you inherit a valuable painting and you wish to sell it, would it be best to sell it in a gallery or at auction? If the latter, would an open or sealed auction be best?[30] Though highly technical and arcane, rational choice theory clearly bears on your dealings with a negotiator like Gandhi—and with suicide bombers and certainly with a nuclear Iran. You need to know: Does the other party reason in a consistent and predictable manner? Is the other side interested in improving their situation in the world? If so, you may be able to come up with various strategies to counteract the other side's strategies. In more technical jargon, you can apply the best techniques of game theory in order to outplay the other side to your advantage.[31] In the case of a nuclear Iran, it may be possible to deter a religiously driven government from escalating a nuclear showdown to the point of no return.

For example, if Gandhi or the Iranians are rational actors, it is conceivable that they are bluffing: Gandhi will take food after four days, and the Iranians will back down when confronted with certain destruction; both are merely sending intimidating signals. Or it is possible that Hamas and Islamic Jihad in the Palestinian territories have not chosen the specific method of suicide terror because they are irrationally bent on self-destruction or because they have found individuals who do not care about dying. One may posit, for instance, that they too are sending certain signals by engaging in seemingly irrational acts or making "risky displays." Risky display is familiar to biologists and anthropologists as an implicitly sensible form of behavior that appears to be suicidal.[32] For instance, when a lion approaches a herd of springboks, or Thompson gazelles, some of the strong young males will spring high up into the air several times in a manner that seems to waste their strength. Why would they do this, and how could they have survived to reproduce? The answer is that they are signaling to the lion that his efforts at a chase will be wasted. If he gets the message (invariably he does), then the spent energy is well worth it. Other animals also engage in risky behavior—male fish approach predators, and even the plumage of some male birds, like the peacock, is risky. The primary benefit of demonstrating indifference to danger is attracting females. The reproductive benefits have clearly outweighed the predation risks (in terms of the evolutionary calculus) because the behavior is common among some species. But what about humans? Does the concept apply to us as well? Indeed, many evolutionary psychologists argue that the young man in the Porsche convertible is demonstrating his ability to waste money and his willingness to risk speed. Such displays, costly as they may be, have proven benefits from an evolutionary point of view—at least in some cultural settings (Southern California, for instance, but probably not Cambridge, Massachusetts).

The risky display program, if not the medium in which it is instantiated (for example, conspicuous consumption versus casino gambling), may extend beyond individuals to groups and organizations, and it may be applied in a tactical manner within military-political contexts. The terror groups may be signaling that they cannot be deterred and that the costs of fighting them will be enormous. A bomb stashed in a bus inside a backpack (the old PLO method) sows terror, but a suicide bomber on the bus signals that the costs of fighting terror may be unacceptably high. What are these costs? For a democracy like Israel (or for the United States), the costs might exceed the bounds of what is regarded as morally justifiable or politically feasible. In other words, in order to defeat its enemies, a nation might have to become a bit too much like them. How many Israeli leaders would consider collective punishment on a Nazi scale? Would they destroy (or massacre) a whole village for the crime of one man and rape all the women in his family? Would they engage in ethnic cleansing and similarly brutal responses? Even the Russians in Chechnya have not gone as far as the Romans or Babylonians might have in the distant past.

Nonetheless, despite all these problems, if the terrorist (or Gandhi) is acting according to some self-interested rationale, then we can figure out a way to better him. That is what game theory is all about.

Among the many game scenarios, the most famous is the prisoner's dilemma.[33] That game is quite similar to the lost object game: if you and your partners in crime are arrested and the police tell you that by implicating your partners you will get time off from prison, what is your best strategy, given that your partners have been offered the same option and only one of you can win? According to Mafia films, the best strategy is to keep your mouth shut, but according to game theory it is to sing like a canary.

All of these games rely on the assumption that both you and your opponent are rational—that is, interested in winning. Winning means maximizing your self-interest, of course. But in the real world, not the ideal world of economists, rationality is not necessarily the most common or even effective strategy. A recent variation on the prisoner's dilemma, called traveler's dilemma, shows this. Consider the following example.[34] Two players report the loss of an identical item to the manager of an airline's lost-and-found. The manager cannot assess the value of the item (some mysterious object you brought from a safari), so he says: "Each of you give me your version of the cost of the item, written on a piece of paper.

Whichever amount is the smallest will be the recognized value; I shall give that person a five-dollar bonus and subtract five from the other. The maximum amount allowed is one hundred dollars." If the purpose of the game is to be the one who gets the largest amount, what should the player write down? According to Basu, a professor of economics at Cornell who invented this game, the vast majority of players who play this game, including experts in game theory, get it wrong. Common amounts given vary from ninety to ninety-nine dollars. The most successful strategy, by far, is to pick the miserably low number of five dollars. Winning rationally requires a willingness to sacrifice profit.[35]

Furthermore, actors are part of larger groups (families, neighborhoods, churches, nations) in which the individual concerns of their self-interest are quickly mixed up with the satisfaction of social rewards. Imagine a woman who has a law degree and has passed the bar, but who gives up a lucrative job in a law firm in order to raise two children. Is this irrational behavior on her part? She is, after all, giving up a great deal of money and perhaps prestige and power. Or imagine someone who donates his kidney to a boy from his congregation, or a professional football player who gives up his career to join the Army Rangers and fight overseas. If rational behavior is about self-interest, either these people are irrational or we have to redefine "self-interest" to include what is regarded as social capital: esteem, approval, honor. Most rational choice theories today do this: they stretch and pull the meaning of self-interest so that even altruistic (or value-inspired) acts of sacrifice can be regarded as ultimately self-interested. For instance, even the football player—the most self-sacrificing of these examples—is acting self-interestedly by defending his country (and family) and enhancing his reputation as a great man.

But obviously we cannot simply argue away the true difference between self-interest and altruism. We do many things solely for our own satisfaction, and once in a while we perform an outstanding act for the benefit of others and against our own interest. To argue that there is no difference between these two situations would eliminate the very meaning of the words "altruism" and "sacrifice." It would also obliterate one of the most important dilemmas in ethics, political science, religion, evolutionary biology, and many other fields—namely, why do people sometimes act simply for others, and when should they? How could altruistic and sacrificing behavior have survived the ruthless evolutionary mechanism that favors the so-called selfish gene—Richard Dawkins's memorable term for the basis of evolution?[36] In other words, if it is the individual organism that bears the burden of survival, how could evolution have tolerated individuals who act truly on behalf of others? Such genes should long ago have disappeared.

Some rational choice theorists argue that the satisfaction people get from acting for the benefit of society may derive from conforming to social values, religious norms or goals, or ethical ideals. These are not based on self-interest, strictly speaking, but they are somehow satisfying (and rational) to pursue.[37] It is also possible, as Amartya Sen has famously proposed, that people settle for enjoying the exercise of freedom, at least the freedom to choose between available options.[38] For example, it may be truly satisfying to be able to choose between potatoes and rice if steak and salmon are not available. In fact, it may be more satisfying merely to have the option of choosing to eat or to fast than to be made to eat a single food type, however delightful. The terrorist who chooses the day and method of his own death may be deriving greater satisfaction from the act of choosing freely than the man who merely survives a dreary and involuntary status quo. That would make the terrorist's choice rational—that is, hedonic (pleasure-oriented) and predictable.

So rationality can encompass a variety of values and types of satisfaction, not just objective "interests." This broader understanding shrinks the gap between social needs and individual desires and enables us to understand that suicidal actors are rationally pursuing some measure of satisfaction. Admittedly, this still sounds unpersuasive: how can the abstract satisfaction of sacrificing oneself for the social good ever match the all-too-concrete pleasures of tasty food, a cool breeze, a loving family, or fulfilling work? The sacrifice seems too much like the complete renunciation of pleasure to be measured in terms of satisfaction. The original gap has just moved from individual versus society, to pleasure versus "satisfaction."

This brings us back to two of the three characteristics of the suicidal aggressor: his religious beliefs and his declarations of joy. Recall that Gandhi promised the Jews who succumbed to Hitler not just the assurance of being right but, so much more compelling, true and lasting happiness. As it turns out, the religion and the joy—which I call the hedonic—are either one and the same thing or two inseparable facts. Only religion offers a compelling way—for all those who take religion seriously—to turn the act of sacrifice into the very source of joy and happiness. Religion is an alchemical hedonic device: it successfully transforms the straw of simple pleasures into the gold of happiness, which is what satisfaction means.

Religion is unique in this respect. True, important philosophers have argued that all human action is geared toward happiness. The French philosopher and theologian Blaise Pascal wrote: "All men seek happiness. There are no exceptions. . . . This is the motive of every act of every man."[39] But this sweeping statement of psychological hedonism has been debunked, and even common sense rejects it. The motive for any given act—choosing bottle-feeding instead of nursing, mowing the lawn or trimming the hedge, replacing the car's generator or its battery—is not necessarily motivated by the search for happiness unless we eviscerate the concrete and conscious intention of actors in real time. Each choice has its own distinct features, and the one we make often yields greater frustration or hassle than would have emerged had we made another choice. It may simply be the case, for example, that the car will not start until the more expensive repair is made. Religion is different. Directly or indirectly, it binds the pursuit of *all* its goals to something it calls "salvation" or "redemption" in a package that is joyful by definition.

Because religion is the most thoroughgoing hedonic institution, it hangs a person's calculation of potential happiness on the hook of a theological formula that strongly favors the group. If you find yourself confronting a player whose strategy is deeply religious (and therefore personally disinterested), your options for success in the game change dramatically. For instance, you could try to persuade the other player that his religion is wrong—that is, that his version is not the correct version of his particular religion—or you could convert him to your own religion. The second option might work (with help from sympathetic religious authorities). But if your opponent is a nuclear Iran, you will probably never win by offering inducements (money, roads, schools) or issuing threats (seven nuclear bombs for each of theirs). Instead, you need to understand the unique (religious) form of hedonic rationality that determines how your opponent plays his game.

This is an exquisitely difficult task unless you understand what "religion" means and how it figures in offering someone deep and lasting happiness. For example, we can decide that Gandhi acted recklessly with his own body because he believed in karma, that some Iranians will act dangerously because they are awaiting the arrival of the twelfth Imam, and that suicide bombers blow themselves up because of a desire for seventy-two virgins in heaven. But none of these conclusions is truly persuasive. Despite being essentially hedonic, religion does not act simply as a metaphysical reward or justification mechanism. Although such beliefs—and even faith in a loving and just God—are central to many religions, understanding the beliefs does not help us understand and eliminate destructive religious behavior. The answer, unfortunately, is far more complex and the dangers we are confronting are far more intractable. Readers of this book will find a solution in the final chapters, but in the process of getting there they will also discover the following:

- Religion is not simply, or even primarily, a matter of belief.
- Pleasure in all its forms, including happiness and joy, is a product of evolution.
- The many aspects of religion that give us our deepest joys have evolved biologically.
- These joyful aspects are by far the most pervasive and the most invisible.
- The destructive aspects of religion are inseparable from the constructive aspects, since both give rise to a profound feeling of joy. Shining a light on the connection between these destructive and constructive aspects can uncouple them. This is the secret to preventing a religious catastrophe.

CHAPTER 2

The Mysteries of Pleasure

As mentioned in the first chapter, Blaise Pascal said that everything we do is geared toward our happiness. The British utilitarian philosophers John Stuart Mill and Jeremy Bentham, in stating that pleasure is the benchmark for everything we consider good, agreed with Pascal. And a distinguished psychologist working in Canada today, Michel Cabanac, has been demonstrating empirically that whether we know it or not, the true motive of our actions is the pursuit of pleasure.[1]

To most readers this sounds like an exaggeration. I confess that I too find it hard to accept. Consider a simple counterexample: When I grade my students' papers, I always take longer to grade the weak ones. I write more comments and make more corrections. Why? Surely it gives me no pleasure to take longer; in fact, I find it rather frustrating. But it is the right thing to do from a professional point of view. Is that a pleasure-driven choice? Not directly. One could make the case that in the long run doing a job well leads to satisfaction and ultimately happiness, but certainly as I dive into a stack of papers for grading, neither pleasure nor satisfaction are at the forefront of my mind. In fact, I would probably make the same choice even if someone assured me that it would not in any way lead to my happiness. At least, I believe this to be the case.

So the philosophers and psychologists may be overstating the case a bit. Still, we do tend to seek happiness, even when a present choice offers up immediate discomfort and frustration in favor of delayed gratification. Moreover, it may be safe to assume that if our actions offered no possibility of either immediate pleasure or ultimate satisfaction, we would probably reconsider our choices. Would we toil away at raising children, or work sixty hours a week, or volunteer for the army, if all we were promised was grief? Not likely. This strong link between our actions and our expectation of happiness is particularly poignant when we contemplate the ways in which religion has become implicated in killing and dying. Nothing seems more tenuous than the connection between extreme violence—be it against others or one's own body—and true happiness. And yet it is often the religious actor who demonstrates precisely such a link.

Consider, for example, the infamous video showing Osama bin Laden and some of his lieutenants celebrating the news of the death and destruction in New York on September 11, 2001. A viewing of the footage reveals that they appear truly happy at the suffering they caused. The same happiness was on display at the Palestinian festival with its celebration of the Sbarro killings, and we see happiness in the comments made by suicide bombers and, sometimes, their relatives. We also read about such happiness in the literature about martyrs and mystics who gladly suffered for a higher purpose.

It is important to emphasize that terrorists and martyrs, like other religious actors, seldom pursue their actions for the *explicit* purpose of attaining pleasure or happiness. Rather, they talk about justice, purity, truth, loyalty, and even heaven, raptures, and love.

Indeed, the satisfaction that attends their accomplishments, or at least their "noble struggle," seems to them merely a *by-product*, just as the minor satisfaction I get from grading conscientiously hardly acts as the true motive for my efforts. But let us consider this: is it possible that in cases of religious violence we assign too much credit to the stated reasons or ideals that consciously guide the actions of these religious actors while ignoring more basic motives (pleasure) that may in fact be decisive?

We are shocked and surprised that anyone would derive pleasure from mass killings, even when he deems it necessary for his broader purposes. No one in the White House cheered (I believe) when Hiroshima and Nagasaki were destroyed by American atomic bombs in August 1945. Where does the pleasure come from when someone is killing for what he considers a noble cause? How do we reconcile his "Islam"—his faith in a peaceful and just religion—and his pleasure in hurting others (or himself)?

I argue in this book that we can understand and disarm the violent and self-destructive actor by analyzing, not his professed reasons for acting, but the pleasure of the act. It is possible to uncouple the actor and the act by removing the pleasure motive, whether or not it is foremost in his mind. How can we do this? We need to see above all else that religion is a preeminently hedonic (pleasure-giving) institution. Despite its complicated theologies and simple faith, its theories of the afterlife and justice, sin, forgiveness, grace, redemption, and all the rest, the basic goal of religious action, as I prove in the next chapter, is the pursuit of what Saint Augustine called supreme and lasting happiness. The terrorist and the martyr are happy in destruction because they are religious actors and death is the means to their end.

The reader may now be curious to know whether I have joined the hedonists cited at the beginning of this chapter—Pascal and the others—who claim that pleasure is behind every action. The answer is no. My thesis is riskier. I am claiming that specifically in the domain of religion (unlike, say, education), pleasure and its nephew happiness are the key motives. Consequently, if we wish to defeat religious terrorists or stop martyrs, we do not go after their ideas and beliefs about God, the afterlife, and justice, but after the pleasure their acts and beliefs produce or are expected to produce.

So what is pleasure and what is happiness? And how can I possibly claim, as indeed I do, that religion is a sophisticated hedonic production machine? That will be the subject of this chapter. Unfortunately, what seems obvious and rather simple—namely, pleasure—happens to be one of the most complex and elusive topics in all of philosophy. So I hope the reader will bear with me as I justify my outrageous claims with a lengthy discussion of pleasure. Let's begin with the following simple example: It's the end of August, which means "two-a-day," the grueling regimen of two full workouts a day, one in the morning and one in the afternoon. The Georgetown campus is still empty except for the football and soccer players who are here to get into shape. It's hot every day, in the nineties, and a water break is the greatest pleasure you can imagine—or even better, a Gatorade break. It's absolute bliss in the middle of a hot and strenuous workout. By the way, have you ever tasted a Gatorade drink with a pasta or seafood dinner in a good restaurant? Absolutely disgusting, but we'll get to that later.

So pleasure seems obvious. We know it when we have it, and some things, such as liquids on a hot day when we're exhausted, are packed with pleasure. This is simple; you don't have to be a philosopher to claim that Gatorade causes pleasure. But what exactly is the pleasure? Is it the taste of the drink? Or its temperature? Is it the sensation of cold liquid pouring down the throat? These would be called sensations—raw perceptions, the data provided by the senses. Is pleasure just sensations? No, clearly not. After all, we feel the pleasure even after we are done gulping, and the pleasure seems to pervade our body in far more mysterious ways than the sensation of cold liquid in the throat. Is it a feeling, like optimism or joy? Are we simply relieved that we won't die in this heat after all? Anyone who has experienced the Gatorade moment, so to speak, may have had some of these feelings, but not necessarily.

When we pause to think about what pleasure truly is, not about what causes it, and try to describe just the pleasure—not the coolness of the drink or the feelings that somehow run parallel with the sensations—most of us are left speechless. I know I am. Because of this quality of pleasure, which some phenomenologists call its "verticality," we often attribute our pleasure to the wrong causes.[2] This error, which I have called the "Prozac effect," following Peter Kramer, will come up throughout this book. As noted earlier, the term refers to a false attribution of our positive feelings to the world outside ourselves, while the true cause is either pharmaceutical or organic but beyond conscious awareness. In chapter 7, the Prozac effect will figure prominently in the way religious individuals, and especially mystics, experience God as real.

Virtually all the great philosophers in history have known that pleasure is confusing and have sought to describe and explain it.[3] I shall oversimplify and divide three millennia of work into four types of explanation and make things even worse by giving them four crude titles:

1. *The ice cream theory:* This theory of pleasure is the one most of us hold about pleasure: it is simple and just a given, like a scoop of ice cream. Some things, like a burning fire, cause pain while others cause pleasure, and the difference is due to the characteristics of objects and their opposite effects on us.

2. *The Greek philosophers' theory:* This theory rejects the simplistic ice cream view and holds that you need to have the right beliefs or judgments about an experience or object before it can give you pleasure. The pleasure itself, like Gene Siskel and Roger Ebert disagreeing about the same film, is a mental event.

3. *The Darwinian theory:* The third theory—the one I shall advocate—proposes that pleasure is the product of evolutionary process and the signal of adaptive success. Humans and many animal species feel pleasure about events or objects their bodies have evolved to enjoy—such as sweet foods—because they are useful.

4. *The smoke-and-mirrors theory:* This theory bears mentioning for the sake of completeness. On this odd theory, pleasure and some other states of consciousness are grand illusions, like the Wizard of Oz. The theory is associated with philosophers such as Ludwig Wittgenstein and Gilbert Ryle and with behaviorists who argue that "positive and negative reinforcements" are real but pleasure and pain are not.[4] Because I have found no discussion of religious psychology that mentions this theory, I will say nothing further about it.

The three theories considered here are large boxes containing numerous distinct approaches to the problem of pleasure. However, they are also radically different; it is hard for the nonspecialist to imagine that philosophers would disagree so dramatically on the nature of pleasure and on what causes it, but they do. I briefly describe here each approach, then state its advantage over the others and its flaws, fatal or otherwise.

Why go to all this trouble? Recall bin Laden in his cave laughing in joy and praising Allah. If he is killing civilians for the glory of God, then he is acting for the sake of pleasure and happiness. God is a proxy for supreme joy, for feeling really good. Some people do horrible things out of a sense of misguided duty, or in panic or anger. These actions do not interest me. Religious actors are driven by a desire for deep and lasting happiness; their religion offers spiritual satisfaction as the incentive for acts of destruction. I believe that in order to stop them we need to understand clearly the nature of pleasure (along with happiness and joy, which are related to pleasure). Only by unraveling the mystery of pleasure and figuring out (in the next chapter) how religion triggers states of happiness can we combat the religious terrorist or stop the martyr from self-destructing.

The Ice Cream Theory of Pleasure

Some have called the ice cream theory of pleasure—or what can also be called the "bottom-up" approach to pleasure—the simple view. In somewhat more technical terms, this view of pleasure claims that "pleasure is a simple kind of momentary experience, the ultimate goodness and motivating power of which are self-evident or observationally obvious."[5]

Such a definition of pleasure has a very long historic pedigree, which means that it is not only obvious but also influential. The Greek school of Hedonism (fourth century BCE) was famous for advancing the notion that pleasure is a momentary but attractive experience. "No pleasure is a bad thing in itself," Epicurus said, although he admitted that pleasures may become complicated owing to the causes that produce them.[6] Nevertheless, on this theory the following, regulated by thoughtful moderation, is true about pleasure: every episode of pleasure is intrinsically good; physical pleasures tend to be more intense, though not superior, to mental pleasures; and the overall value of a whole life is the sum of all its pleasures after the subtraction of its pains.[7] After all, if you eat too much ice cream, your assessment of the overall pleasure will diminish. The good and desirable life depends on the carefully modulated accumulation of pleasure.

Hedonism here runs into ethics. What feels good is good. Modern philosophers, especially in Britain (from John Locke to Jeremy Bentham), famously agreed with the hedonistic ethical proposition that the good life is the pleasurable one. Locke put it in the following terms: "Things, then, are good or evil, only in reference to pleasure and pain."[8] This claim, under the pen of Bentham, would later become the utilitarian philosophy that living the ethical life—knowing the right thing to do in specific circumstances—depends on maximizing pleasure and minimizing displeasure: "It is for [pleasure and pain] alone to point out what we ought to do, as well as to determine what we shall do."[9]

Obviously these views have become very influential in our culture and even in our sciences. In many scientific fields, like economics and psychology, rationality itself has come to be defined as the effort to maximize one's interests, which means accumulating pleasure or happiness.[10] We shall get to this important point later when we discuss methods for minimizing the pleasures of the religious actor. The ice cream view of pleasure meets the standard of common sense and simplicity, which says a great deal. However, it suffers from a number of severe limitations.

Problems with the Ice Cream Theory

This simple picture of pleasure, the one most of us accept as a matter of common sense, raises several puzzling questions. Chief among these questions are ethical concerns, and they fly in the face of the Hedonists, Locke, and the utilitarian thinkers who argued that the good life is constituted by the sum of one's pleasures. Imagine a serial killer who manages to murder twenty-three people. He anticipates each killing with rising relish, then savors the killing for months afterward. In between is the gruesome act, which is an ecstatic moment for the killer. This killer's life—a life dominated by his many killings—is cut short at age forty by execution. Can we say that he lived a morally good life? Certainly not. But can we even say, adding up his moments of pleasure, that this man lived a "satisfying" or happy life by his own reckoning? Those who question whether such a life can be described as fulfilling do not doubt that the killer enjoyed his crimes; what they doubt is whether his pleasure—or any pleasure—can be the foundation of a fulfilling life. We are just too intimately wed to the ideal (fathered by Plato) that happiness requires a certain level of virtue.

Aside from such ethical consequences, other issues associated with pleasure are just as compelling. Among the oldest and most intriguing is the puzzle of masochism: How is it possible to enjoy a pain such as humiliation, bondage, or whipping? If pleasure is intrinsically hedonic—that is, it feels good and is desirable—how can it be brought about by pain? Examples of masochism abound: "I was still beating him, putting every ounce of strength into my blows . . . he was mutely savoring the interior stirrings of delight."[11] The hedonist who insists that pleasure is a "simple idea" (Locke's term) and a single one that varies in its causes and its level (quantity) but not its feel cannot easily brush aside the challenge that masochism presents.

A similar problem arises with respect to complex pleasures—spicy food, bitter coffee, beer, a sauna, a roller coaster: if pleasure is an "elementary idea," how can it consist of unpleasant or frightening sensations? Related to this is the question of learning. Many of the pleasures we enjoy, such as those just mentioned, are learned. If that is the case, how do children acquire new pleasures (spiciness, hard work, reading a long book) out of activities that start out as neutral or unpleasant experiences? At what moment does the aversive experience become pleasant, and at what moment does it consist of both pleasure and displeasure?

Still more incisive doubts arise concerning simple (ice cream) pleasure. The most telling can be illustrated in the following common example: Imagine being stroked on your arm by a lover and think about that feeling of pleasure. Now replace your lover with an imagined rapist who has you strapped down, or perhaps a torturer who is a sexual sadist. Does the same feeling of pleasure persist when this hand touches you? If this example does not persuade you, try this one: Imagine getting invited to your boss's house for a July Fourth barbeque where, instead of hot dogs, you are served live, wriggling maggots by your host himself. He has recently returned from New Guinea, where he learned to love nutritious grubs, and has now decided to bond his employees by testing their resolve to eat them as well. So go ahead—bite into that fat white worm! Would it help if the grub were cooked, or if you were told that it actually tastes better than a hot dog? Many Germans, incidentally, react to garlic as most Americans would to a worm—and this attitude spills over to those who eat garlic.

If eating is an intrinsically enjoyable sensory experience, based on taste, succulence, and aroma—as is a light stroke on the skin of the arm—why can't we visualize these examples as pleasurable? Can it be that to enjoy something sensual we also need to have the right frame of mind, with the right beliefs and attitudes? How deeply does our frame of mind influence the pleasure we take for granted as simple and momentary?

The Greek Philosophers' Theory of Pleasure

THE GREEK PHILOSOPHERS' THEORY of pleasure has been the most influential approach in the West and has particularly saturated religious thought. Jewish, Christian, and Muslim theologians would not be able to think about God or about pleasure without it, and so, despite the fact that it is essentially incorrect, I devote a good bit of space to it here.

Most thinkers, going back to Plato, have rejected the notion that pleasure is simply there, like the redness of a rose. Instead, if pleasure is to enter our consciousness, some evaluation—a fairly sophisticated mental calculation—has to be brought to bear. You might almost say that pleasure is a judgment ("boy, this is good!") *about* what we see and taste.

Much of the philosophical opposition to hedonism (in both its psychological and moral forms) does not go so far as to reject simple pleasure altogether. Instead, it insists that a variety of criteria, "top-down" notions or conditions, must kick into gear in order for a sensation to be perceived as pleasurable. Whether imported or intrinsic to pleasure itself, these criteria make pleasure a far more complex experience than we first assume.

Plato may have been the first to raise this point when he argued, in the mouth of his teacher Socrates, that pleasure is not simply an intrinsically enjoyable sensation. Instead, it is the awareness of the transition from a state of depletion to a state of replenishment, from lacking something to having it. Pain is the opposite—the transition from having something to losing it. There is a set-point, a perfect balance or equilibrium among the physical and mental elements of our existence. Thirst and hunger represent depletions, while drinking and eating act as replenishments. The consciousness of the change from one state to the other is pleasure or displeasure. Additionally, drinking or eating beyond the point of equilibrium (satiation), regardless of how tasty the food or beverage may be, decreases pleasure.

The obvious problem is that not all pleasures act in such a way. We do not enjoy a movie or a baseball game because something is lacking. We do not, in fact, prefer soda pop to water or juice because it replenishes us more effectively. Many transitions, such as aging, are too gradual to be perceived in the present moment of awareness. Finally, we seem to enjoy being home after work more than the commute. Clearly, something besides transition must be involved in the creation of pleasure.

A generation later, Aristotle added further insights to Plato's discovery. In some of his earlier work, he simply elaborates Plato's argument: "Let it be laid down that pleasure is a certain process of change in the soul, viz. a sudden and perceptible attainment of the natural state which belongs to it, and distress the opposite."[12] The added implication here is that the set-point is specific to the species and to each individual—it is not some cosmic abstraction ("balance"). For example, the mathematician attains pleasure by reaching the full potential of his reasoning abilities, while the athlete attains the same from the full development of his body. In other words, the real pleasure is the perfect *actualization* of the nature of one's species—with the highest pleasure owing its existence to the perfection of life as seen from God's objective point of view. A perfectly just jurist or an artist who creates a perfect statue approaches such a standard.

Think of a man struggling in the desert, exhausted and thirsty, who chances upon an oasis. He falls to his knees by a clear fresh spring of cool water and begins to drink. Plato explains that the drinking, as it causes a change from dehydration to saturation, is perceived as the pleasure. But now the man has cooled down and restored his fluids, and he is resting under a tree. For Aristotle, that is the higher pleasure, the enjoyment of the state of saturation, or perfect function relative to the proper way of being. While Plato emphasizes the perception of change, his younger contemporary sees the perfected state—call it self-actualization—as the source of pleasure.[13]

Both Plato and Aristotle (and the Hedonists as well) recognized a sharp distinction between soul and body and ranked them accordingly. Naturally, they also separated different domains of pleasure—sensory, emotional, intellectual—as most people still do today. The perfect functioning of the organism in digestion is a source of *physical* pleasure, while the satisfaction of figuring out how to program the DVD recorder is *mental*. The Greeks—and following them, the Christian theologians—insisted that the pleasure of the philosopher who pursues truth is far superior to the pleasure of the athlete who wins a contest. According to Aristotle, the "species" of the actor (gender, profession, religion, and so forth) determines the proper goal of his efforts. When that goal (*telos*) is fully realized—when the nature of the actor flowers from potential to actualized—true joy becomes manifest. Thus, although there are many different pleasures, they all depend on the same two basic mechanisms: perception of change and the realization of one's natural faculties. We can call the first one *dynamic pleasure* and the second *fulfillment pleasure*.

Both Greek pleasure criteria, along with the body-mind split they required, were profoundly influential in Christian history and remain so today. Saint Augustine, the most important early theologian, puts it in these words: "In eating and drinking we restore that which daily falls away from our bodies, until such time as you destroy food and the stomach, when you will slay our need with wondrous satiety."[14] The bodily change itself, from hunger to satiety, is the source of pleasure, but that pleasure is addictive, Augustine explains. It causes us to seesaw from one state to another: from hunger comes the desire to eat, followed by eating and back again—endlessly. "The very transition contains for me an insidious trap of uncontrolled desire." Moreover, we insist on having our food prepared in an increasingly appetizing manner in order to arouse the palate, which has been sated by overindulgence.

In the jargon of contemporary psychology, this is known as the hedonic treadmill. Obeying the constant urge to replenish the body's many needs, but jaded by frequent indulgence (habituation), we keep reaching for greater and greater stimulation—more purchases, louder music. To avoid the body's dependence on such cycles and bring to an end its addictions, Augustine insists that a higher purpose must be found—something constant. And so, to control this addiction, Augustine imagines a new purpose behind eating: health. "I should approach the eating of food like the taking of medicine."[15] But Augustine knows that we easily deceive ourselves with such fine purposes and that we either become too severe on ourselves or we simply forget when to stop eating. A higher purpose than even health must be found if joy is to be sustained without falling into the traps of either the hedonic treadmill or overly severe self-deprivation. Only God can provide such a goal, which, once obtained, removes the body's temporary pleasures and desires: "And the blessed life is to rejoice in the Truth; for this is rejoicing in you who are the Truth."[16]

Augustine proposed that acting according to an Ultimate Goal is the cure for the pleasures of change. And that it is also the source of a different kind of pleasure (Supreme Joy). These ideas are far from archaic. Christian writing today is profoundly hedonic in this Augustinian sense. Rick Warren's bestselling *The Purpose Driven Life* is a book about joy, which situates pleasure in the same ultimate goal—in God: "When we give God enjoyment, our own hearts are filled with joy."[17]

It is important to note that for Aristotle, this joy (fulfillment pleasure) is nothing like divine grace—it is a mechanism. Indeed, one need not be a believer in God to obtain it. The marine is happy to serve his unit, and the hospital volunteer derives pleasure from serving her community. The pleasure comes about neither from sensory stimulation nor from a divine agent. It comes from the mental appraisal of a larger agenda.

According to the Greek philosophers' theory of pleasure—based in awareness of transition and a search for a higher purpose—no sensory experience is ever intrinsically painful or pleasurable. The elementary units of any experience—call them "phenomena"—and the basic qualities of sensations (the redness of a rose, the tanginess of lemonade) are neutral, neither pleasant nor painful. Something else is needed. We need to answer the following questions: Does this experience serve a goal? Is such a goal any closer now than it was before? For example, when your muscles are sore, you can evaluate that sensation in any number of ways. Are you carrying a flu virus? Or did you have a soccer practice yesterday? You write a check for $5,000: is it for a wedding or a funeral? You're drinking something bitter: is it cold or hot? Everything needs to be appraised in order to be felt as pleasant or unpleasant.[18]

Problems with the Greek Philosophers' Theory

The top-down theories of pleasure generate a new set of questions that philosophers must resolve. If pleasure involves evaluating a broader context, how can newborn infants—or indeed, animals—have pleasures? Clearly they do not possess the mind of a Greek philosopher. To say that their goals and their evaluations are implicit or nonconscious only confuses the issue: How can a baby who likes a lullaby "evaluate" the song in any meaningful way, implicit or otherwise? She just likes it, no? Furthermore, why do so many of us have similar pleasures? Undoubtedly, we value different goals and different ways of assessing our progress—so how could our pleasures be so widely shared? The brain surgeon likes to excel and be recognized as much as does the athlete, who in turn enjoys word puzzles as much as the professor. And everyone likes chocolate. Furthermore, how do we account for guilty pleasures? You sign up with a weight-watching group because you positively yearn to lose forty pounds, but ice cream still gives you as much pleasure as ever—perhaps even more.

If the problem with simple pleasure is that it is too simple—that it is atomistic and incoherent—the problem with different versions of intentional pleasure (the Greek philosophers' theory) is that they are too mental and too metaphysical—certainly for infants and pets. They require a rational subject who imposes order on her world via thoughts and values that are far removed from the simple experience. This feels too contrived, like hearing about someone who only enjoys a dish after memorizing the recipe.

The Darwinian Theory of Pleasure

A THIRD APPROACH WOULD have to take into account the advantages and shortfalls of both the ice cream theory and the Greek philosophers' theory. Pleasure, unlike trees or furniture, does not simply occupy the space in which humans and animals move. Some standard for evaluating the interaction between the actor and the environment must first do its work. However, this standard must also be built-in, yet verifiable; universal, covering animals, infants, and philosophers alike; and accessible in every case of experienced pleasure. According to Daniel Nettle and other new theorists of pleasure and happiness, the universal standard is not some broad notion such as the good life, well-being, self-actualization, or the perfection of our potential.[19] All of these are cultural facts—lofty ideas that would clearly exclude children and animals from the experience of joy or even pleasure. Instead, the standard that turns sensations into pleasures and pains is biological; evolution has programmed us, like many other organisms, to have a set of goals that define what is good and what is bad, what is pleasurable and what is painful. The name for this evolutionary approach to pleasure is *functionalism*.[20]

Evolutionary goals can be stated in disarmingly simple terms. Survival, reproduction, and prosperity within a given environment, including a social one, would cover much of this topic. An act, a sensation, or an emotion gives the actor—even an animal—pleasure when it contributes to survival and prosperity—when, in a word, it is *adaptive*. The details, of course, are worked out in a variety of ways. For example, bears get pleasure from honey because they need the calories; bees do not find honey sweet (or pleasurable) in that way at all.[21] Humans too find fatty and sweet foods pleasurable to eat because the nutritional value of such foods was essential for human survival in early environments. To put the matter in slightly different terms, pleasure is the reward for adapting to situations in which there is a choice to be made: to eat or stop eating, to mate or not to mate, to run or stay in place, to attack or not to attack. When our ancestors made the correct (adaptive) choice based on seeking pleasure and avoiding pain, they survived to reproduce. Pleasure is their legacy to us.[22]

Adaptive Scripts

We need to avoid the impression that just because adaptation is about survival, pleasure is a simple matter. To begin with, the word "adaptation" applies to events that range in duration from millions of years to milliseconds. For example, the emergence of new organs in new environments (eyes, limbs, brain) takes many generations, and although they are adaptive, no pleasure is associated with such changes. On the other end of the scale, travelers to Denver need to adapt to the elevation there, and more quickly yet, our pupils need to adapt to light or darkness as we move from indoors to outdoors or vice versa. No pleasure there either. What we are talking about when we say that pleasure reflects adaptive choices is known as "behavioral adaptation." This is what evolutionary psychologists study, whether they work on human or animal populations (behavioral ethology). Animals with sufficiently developed brains make correct decisions by following environmental cues and responding (adapting) in ways that lead to pleasure.[23]

Even behavioral adaptation, however, is complex.[24] At the very least, it consists of three elements: prior state, novelty, and response. These can be called the basic components of the adaptive script. Let's look at the example that opened this chapter, the thirsty runner. Thirst begins as a cellular event (loss of fluids and minerals) that does not reach the awareness of the runner until after a certain threshold has been crossed. The brain registers that a loss of balance has occurred and that restoration of the original state (homeostasis) requires that an action take place, which requires, in turn, conscious awareness or attention. What the runner becomes aware of is novelty, experienced as some discomfort. He easily assesses the sensation of thirst and stops for some Gatorade or water. Most runners are not thinking about their cellular depletion but about reducing thirst—feeling good. The water or Gatorade feels great going down because drinking a fluid represents the correct response to the novel situation.[25] I do not know how a gulp of whiskey would feel; I have always been amazed when watching old westerns to see thirsty cowboys quenching their thirst with a whiskey at the saloon.

The human brain has evolved to enjoy the things that were best for our ancestors in the Pleistocene environments. That is why we love honey-roasted peanuts so much: fat, sweet, and salty, this is the supreme evolutionary snack! But the feeling of pleasure is produced at the brain—not in the cells that return to their prior state (homeostatic restoration) throughout the body. We owe the pleasure to the neurochemicals, such as dopamine and beta-endorphins, that are saturating the appropriate circuits in the brain in response to an assessment that adaptive goals have been set and met. Unfortunately, just like us, our ancestors rarely found themselves in situations that involved single goals or action plans. Several scripts had to compete for their attention: eat or play, eat here or somewhere else, eat alone or share, seduce this woman or preserve the friendship of her mate, take a nap or go swimming. And of course, the choices seldom involved just two options; to any set of choices several others could be added—take your child along, defecate first, treat the minor cut on your forearm. Some goals were natural, whereas others involved cultural choices. (What do you do with your little boy when you have decided to sneak off with a chunk of meat?) Cultural choices often involve learned action plans, and the resulting pleasures are learned, like acquired tastes.[26] The fact of adaptive scripts competing with learned pleasures results in many of the pleasure paradoxes we encounter today. Here is one example.

Scientists now know that chocolate contains substances such as phenylethylamine that trigger the production of dopamine in the regions of the limbic system that regulate such moods as euphoria and attraction. Other chemicals increase the levels of the neurotransmitter serotonin in the brain. Chocolate, then, should be all good all the time. But it is not. My father, whose mother had persuaded him that chocolate would ruin his teeth, always hated it, and I usually start hating chocolate myself after four squares (okay, six). If brain chemistry causes pleasure, why does it end so quickly, and why is pleasure so vulnerable to suggestion or the power of learning? Or more dramatically yet, why do so many people enjoy behaviors, such as smoking and gambling, that appear to be maladaptive? How can a person come to enjoy self-destruction or martyrdom? To answer these important questions, we need to look more closely at the details of adaptation.

As noted earlier, the basic adaptive script looks roughly like this:

Novelty ➡ Response ➡ Restoration

In reality, the subsidiary steps between the three main steps are just as important. A more realistic version would look a bit like this:

Novelty ➡ Stress ➡ Attention ➡ Evaluation ➡ Seeking/
Avoiding Response ➡ Solution ➡ Balance ➡ Mastery ➡
Exploration ➡ Novelty

In this more textured version, affect (pleasure/pain) has a far more nuanced role than just restoration or disruption (recall Plato). Pleasure, in a large variety of ways, informs the process as a whole. For example, paying attention (the third step) can be highly pleasurable, as we shall see in later chapters when we discuss mysticism. Focused and deep attention, which excludes distractions, uses brain centers and chemicals associated with deeply satisfying states of consciousness. Or, to give another example, when differing agendas (have sex or take a nap) compete for immediate response, pleasure may act as a quick way of evaluating (fourth step) which course of action to pursue. Some steps in this complicated process involve learning, which can produce the distinct pleasure of removing the stress of a new situation. Mastery is the perfect example of this. This step in what I have called the adaptive script includes both achievement of a difficult task and, related to that, the capacity to delay the need for immediate gratification. The successful hunter must master his fear and fatigue. He must learn difficult skills that require patience, concentration, and other qualities that depend on the ability to postpone quick resolution to perceived needs (hunger, sleepiness, and so on).

Finally, solving the initial problem (hunger, fatigue, response to danger) and gaining mastery over the world and over oneself often leads to exploration pleasure. This is the playful and joyful behavior that can range from the play of kittens and children at recess through human art and scientific exploration and all the way to mystical raptures. At its best, such behavior is not merely a response to a perceived need that must be resolved, but the manifestation of an internal impulse toward novelty that has been untethered from mere necessity.

Clearly, the various steps of the basic survival script involve the full range of psychological abilities: emotions, perceptions, cognitions, and even aesthetic assessments. We all know that the solution to a perceived need, such as hunger, can be delayed—and the pleasure it gives thus stretched and deepened—if we learn how to slow down the rush to satiation or if some form of novelty is embedded within the process.[27] For example, playing with your food can make it taste better longer. My cat seems to know this, because she plays with her cat food, but only after the first few pellets. She also plays with flies and bugs—creatures that might have been food if she were not so well fed. And of course, we too play with our food, if not quite in the same way. We vary it, spice it, alternate bites of different foods, and sip wine or carbonated water while we eat. We delay biological homeostasis by lengthening the period between seeking and satiation or by adding other forms of novelty, such as conflicting tastes or spiciness.[28] We become creative and come to expect ever more sophisticated pleasures.

This indicates three things:

1. The steps of the script are usually sequential—in principle at least—but they are not inflexible. The steps can be repeated or moved around a bit, like themes in a musical score.

2. While pleasure seems to be associated with the solution step, this is not entirely accurate. As a matter of fact, pleasure can accompany the attention phase, the evaluation and seeking phases, and the various aspects of exploration that follow the solution. This variability is possible because the basic script

contains subsidiary ones, with their own threefold cycles. Each pleasure, however, has a different feel.

3. The manipulations of the adaptive cycle, such as the delay of satiation, are often cultural creations (cuisine, art, music), and the mastery pleasures they produce are often complicated and learned (discussed in detail in chapter 4).

The importance of the complicated evolutionary-psychological script for the often surprising experience of pleasure can be illustrated by looking at the example of sex. It is common to think that we enjoy sex as an evolutionary reward for reproducing. It is clear from the vast majority of species, however, that reproduction does not have to include a fun component. For humans and a few types of apes, sex feels good because of the interplay of several factors, including hidden ovulation, the long infancy of the offspring (requiring two parents), and the economic exigencies of groups. According to Jared Diamond, the pleasure of sex involves primarily the reproductive strategy of females, who evolved to conceal their ovulation within promiscuous societies. Hidden ovulation eliminated the father's (and mother's) certainty about the identity of the genetic sperm donor. It thus demanded the father's loyalty as a way of ensuring his mate's monogamy. This development also provided help for the mother with parenting. Recreational sex and its attendant pleasures emerged when even the female did not know when she was fertile, and it acted as a reward for jealous monogamy, lasting for several years after reproduction.

This type of biological analysis helps us uncover the hidden scripts that play out with sexual pleasure and explain why sexual pleasure can be so mystifying. For example, we can better understand the so-called seven-year itch: the pleasure of the monogamous relationship begins to wear thin at about the time the child becomes independent and the presence of both parents is no longer necessary.[29]

Conclusion: Sordid Pleasure

THE EVOLUTIONARY/ADAPTIVE APPROACH TO pleasure requires that we modify the usual typologies. Most people, following the Greeks and the Christian thinkers, divide pleasures into the physical (including food and sex), the emotional, the intellectual, and the spiritual. It is the context that gets the nod—whether the object we have in mind when we feel good or the apparent cause of our pleasure—and this is what I have called the "Prozac effect": the incorrect attribution of internal feelings to the outside world. The true causes of pleasure are the neurological and neurochemical events taking place in the context of appropriate responses to adaptive choices. The three main types of pleasure, on this view, are novelty-replenishment, mastery, and exploration.

So let's return to the smoker and the gambler. If pleasure has something to do with adaptive success, why do people enjoy self-destructing? The twin answer is competing scripts and learned pleasure. If we can learn to enjoy Shakespeare's *Macbeth*, we can learn to enjoy very complicated things indeed. That does not mean that hating Lady Macbeth serves an evolutionary purpose, but only that the brain circuits that evolved to experience certain emotions in given situations are extremely flexible. One can use them—the technical term for this is *preadaptation*—for other purposes, which are set by various cultural agendas.[30] For example, the gambler utilizes the brain's arousal states that are activated by uncertain decisions involving risk, such as hunting. The new cultural agendas—

even something as potentially destructive as gambling—represent alternative adaptive scripts and can give us other sorts of pleasure.

Why is Osama bin Laden so happy in his cave? Ask him and he will tell you that, in the name of God, he has struck Satan with a mighty blow. We might experience the same joy if we found out that bin Laden had suddenly died of syphilis. But the cause he attributes to his pleasure—his conscious script—is just one option. I am inclined to think that there are hidden scripts at work for which his sordid theology is merely the proxy. For example, bin Laden is seeking to "purify" the sacred land of Saudi Arabia of its defiling occupiers, namely, American troops. Behind the language of purity (see chapter 5) is an ancient way of reckoning group boundaries: he is fighting for the integrity of his native group, which is a very primitive and compelling motivation for killing (and dying). This type of group-bond precedes and transcends Islam—it is, in fact, tribal. The visceral connection to the group (and its territory) produces states of joy in actions that fortify the inside against invaders from the outside. Osama's pleasure may owe its intensity to a script of which he is only dimly aware, if at all.[31] To make men like him miserable, we must move beyond their stated theology, their talk of Allah and Muhammad, and uncover the additional scripts. This does not mean that religion can avoid being implicated in terrorism or in self-destructive violence, but that the picture has many layers.

CHAPTER 3

The Varieties of Religious Pleasure

In the mid-1950s, two sociologists, Leon Festinger and Henry Riecken, published an account of a middle-aged woman named Marian Keech who lived at 847 West School Street in Lake City, Minnesota, and who "received" a message that the world was coming to an abrupt and catastrophic end on December 21, 1954 or 1955. A flying saucer was to arrive at her house and save her, along with a small group of followers, just before the end. That last night, Mrs. Keech and the core of her followers, all of whom had quit their jobs and disposed of their belongings, sat up praying and awaiting the flying saucer. As the night progressed and nothing happened, the unease of the group grew. But just before dawn another message arrived for Mrs. Keech, who announced: "The little group, sitting all night long, has spread so much light that God has saved the world from destruction."[1] The discomfort of an incorrect prophecy and, presumably, the fear of a great destruction gave way to the exhilarating joy of a new prophecy. This second prophecy validated the first by contradicting it and had the additional benefit of sparing the world. The hard-core members who had sacrificed their properties and jobs did not become disenchanted; in fact, their faith grew, along with their devotion to Mrs. Keech. With redoubled commitment, they hit the streets to proclaim her spiritual powers. Meanwhile, those followers who had entertained some doubts and stayed in their own homes lost all interest.

Why is it that those who by all accounts had given up the most were also the ones who derived the greatest emotional satisfaction from the failed prophecy that night and clung hardest to their beliefs? According to psychologists like Carol Tavris and Elliot Aronson, this is a classic example of cognitive dissonance. People cannot easily accept the fact that they have been easily duped, or that a decision they have made is irrational. Without consciously deciding to do so, they hold on to their original conviction with far greater zeal. They also report a far higher level of satisfaction in their choice. The psychological dynamic at work is avoidance of the mental discomfort of two (or more) conflicting beliefs ("I am a reasonable person" and "That was a really dumb thing to do"). According to anthropologists such as E. E. Evans-Pritchard, who had studied East African magical rituals, this symptom is known as "secondary elaboration": it is designed to explain, in various ways, why both predictions and magical rituals fail to deliver but we must continue to believe them anyway.[2]

Festinger's cognitive dissonance and Evans-Pritchard's secondary elaboration are scientific concepts. They generate predictions about human behavior that can be tested empirically. Although both of these researchers studied groups and their results apply to culture, there is a real psychological and, indeed, biological mechanism at work in the dynamics they describe. Anytime a general rule applies across distant cultures, it seems to explain something basic about human nature. This case points to a basic human need for a psychologically consistent image of oneself and of one's circumstances and a tendency to experience strong discomfort—mental pain—when contradictions surface. An effective mechanism for generating willed blindness kicks into gear to reduce pain and increase satisfaction, with vast cultural ramifications.

This is just one example of a biological theory of religion, or religious thinking; there are others, as we shall see. But if religious behavior can generate biological explanations, it must be subject to evolutionary science—that is, it must be shown to be adaptive in some way. In this case, self-deception yields irrational beliefs but strengthens social ties.[3] We might argue that the very persistence and universality of religion attests to its biological usefulness to some extent, perhaps as a socially adaptive institution.

Readers will begin to object that Mrs. Keech of Lake City and Evans-Pritchard's Zande (East African) magicians have nothing to do with Christianity, Judaism, and all the other "higher" religions and that their practices are merely superstitious. But Christianity also began with prophecies and anticipation; the first generations awaited the coming end with bated breath. Many Christians still do, although not to such a degree that they fail, in the meantime, to go to restaurants, enjoy movies, and even read record numbers of books by Jerry Jenkins and Tim LaHaye about the coming end of the world. Judaism also features a coming messiah (except for Reform Jews, the joke goes, who know he will never come). As a result, Judaism too has seen its share of false prophets and messiahs, most notoriously Sabbetai Tzvi, who died in 1676.[4] Messianic expectation is central to all three Western monotheistic religions; the quirky cold war variety from Lake City simply took the short view on an extended theme.

The psychological mechanism reaches beyond failed prophecy and beyond cognitive dissonance. It goes something like this: the more you invest in your belief, the more you stick to it and the deeper the satisfaction you get from it. Consider rites of passage, those nearly universal rituals around the world in which young men and women undergo painful tests in order to become adults. True, the Jewish bar mitzvah and Christian confirmation rites no longer insist on physical pain, but in many cultures around the world pain is still central. And some rituals of initiation in the West (Marines, fraternities, secret societies, cults like the Moonies or Hare Krishna) still do require discomfort and sacrifice. The harder the test for joining a group, the more loyalty that group generates toward its principles and the deeper the sense of accomplishment for being a member.[5] It hardly matters if the group engages in immoral behavior, and it matters even less that membership can impose self-destructive behavior. The feeling of satisfaction—a form of pleasure—points to the biological and psychological mechanism at work in the need to belong.

Skeptical readers, even those who acknowledge that evolution may apply to human societies, will insist that the violent and destructive aspects of religion represent the exception—an evolutionary aberration. They would point to other violent and self-destructive behaviors as evolutionary aberrations as well. Consider gambling or watching cage fights. Both of these forms of behavior, with potentially destructive consequences, owe their existence to positive evolutionary tools that emerged in the Pleistocene age when the modern human brain evolved. The thrill of the hunt, including its risks and uncertainties, along with the excitement of the kill were vital to survival in that era. Those creatures that had developed the brain circuits and neurochemicals to reinforce violence and risk-taking reproduced successfully and passed those traits on. Today we still get a thrill from risky behaviors, even if our survival does not depend on them. We capitalize on the hardware that is already in place, although this can lead to addiction and to antisocial behavior.[6] Destructive and irrational religious behavior may be similar: it is a phenomenon that merely exploits preexisting brain hardware—in this case, beneficial religious beliefs and institutions.

In this chapter, I show that the same biological and psychological rule applies to both the most sublime and the most self-destructive aspects of religion. The same force that has made religion a profoundly important human institution has also made it a source of threat. The suicide bomber, the martyr, the mystic, and the prophet are members of the same human family: *homo religiosus*.

What Is Religion?

A Portuguese woman stands in the Jordan River, a mile or so from the point where it emerges out of the Sea of Galilee. She is dressed in a white robe, and the water comes up to her stomach. Every now and then she dips completely. The day is hot, but she is cool. Upstream some Israeli kids are splashing in the murky water. Maria is happy. Fingering a crucifix on a necklace, she prays for the health of her husband, who stayed at home in Lisbon. From a card, she recites the prayer of the first luminous mystery—the baptism of Jesus in the Jordan River. She repeats the Hail Mary with precision:

> Hail Mary, full of Grace, the Lord is with thee. Blessed are thou among women, and blessed is the fruit of thy womb Jesus. Holy Mary, Mother of God, pray for us sinners now and at the hour of our death. Amen.

Having brought a letter from her husband, with a prayer for Mary, Maria lets it float away on the water. It sinks down and settles among hundreds of other letters, cards, and photographs. After fifteen minutes, she slowly climbs out of the water and takes her towel and dry clothes into the changing room. Some time later, purse in hand, she walks into the souvenir shop and buys a few packets of Ahava soap from the Dead Sea.

Can anyone deny that Maria from Lisbon has undergone a religious experience or that her joy in the river is a religious feeling? There's the pilgrimage, the baptism, the prayers, the faith in God and in the Holy Virgin. The letter in the water looks like magic, perhaps a bit like the Jewish notes in the cracks of the Wailing Wall, but she may not actually believe that it was even necessary; after all, she prays to Saint Anthony for her husband's health. And of course, like any tourist who's there to see the sights, she came in an air-conditioned bus and paid $50 for the special soap. But these last two observations are trivial—the heart of the matter is clearly religious.

A Lutheran observer might discount most of Maria's activities this day but acknowledge that her faith in God is not only religious but also true. A Muslim might dispute whether her faith is true, but not the fact that it is essential for a religious person. A Catholic would approve of everything—the ritual, the liturgy, the faith in miracles. A Jew or a Hindu would give the nod to the rituals (the bathing and the pilgrimage). Richard Dawkins would define all of this as religious because it is all equally nonsensical, and Woody Allen would probably make a crack about bathing in muck and buying ridiculously priced soaps.

Academic scholars of religion, not surprisingly, cannot agree on what constitutes religion in this case or in any other. Is it the belief (say, in God)? Imagine that Maria had no such faith, although she wished she had. We are finding out now that even Mother Teresa was racked with doubts about God during her years in Calcutta.

Would Maria's experience still be religious if she doubted that God existed or if she were deeply agnostic? A Buddhist pilgrim at Bodhgaya in India visits the place of Buddha's enlightenment—no God there. So, is it the ritual that makes this type of Buddhist practice religious? Does the ritual have to evoke feelings (of joy, exhilaration, guilt), or can it be rote?[7] Perhaps it is Maria's companionship with the other pilgrims. When we speak about "religion" as a distinct phenomenon—say, like art or sports—what exactly does the word point to in the world of religious practitioners? Every possibility has been explored by some experts and rejected by others for a variety of reasons. My professor at Harvard decades ago, Wilfred Cantwell Smith, insisted that the term "religion" is useless and should be avoided altogether.[8]

I am aware that many people may consider this hairsplitting. Religion, like an orgasm, is something that you know when you have it. If Maria would not bother with a definition, why should we? The reason is simple: I argue that religion has evolved naturally and that it serves a natural function. As such, it affects much of what takes place in the world today, including suicide bombing. But what is the "it" here? Clearly not everything that falls under the heading of religion qualifies for the same biological explanation. If, by religion, we mean a special feeling, such as mystery, awe, or elevation, that is one matter; if we mean a church or a congregation, that is another matter altogether. And to make things even more complicated, the true motive of human action is often hidden behind what people think about their religion. Like those who suffer from cognitive dissonance, religious individuals can be blind to what moves them! Here is one example.

In an article in the *New York Times* on June 22, 2007, Jane Perlez interviewed a Somali immigrant in London. Hodo Muse was explaining to the journalist why she preferred to wear the *niqab*—

the head-to-toe garb that Muslim women wear in traditional society. "Wearing the *niqab*," she said, "means that you will get a good grade and go to paradise. Every day people are giving me dirty looks for wearing it, but when you wear something for God you get a boost." This is a perfect example of religious pleasure—acting for God and getting satisfaction. But imagine, for the sake of argument, that Hodo had said, as other *niqab* wearers do, that she wears it as a mark of pride in her own religion. Or perhaps wearing it was an act of defiance against the Londoner's sense of superiority and an assertive rejection of the customary meekness that immigrants tend to show. Or she might have been uncomfortable with the stares of men—she might have grown up in a culture where men regard women as sexual objects. Indeed, it is possible that all of these motives are at work at the same time. If so, it would be appropriate to propose that while Hodo thinks that her special thrill is due to obeying the will of God, in fact it is caused by other factors. Her assumption about the source of her own rejoicing is merely a theory—a folk theory.

So, what is religion? It will probably be enough to say that religion is any combination of some or all of the things that take place around Maria in the Jordan River, or even just one of those things. There may be some belief in its object (God, the afterlife, karma) and a subjective dimension (reverence, awe, joy); there may be a ritual, a scripture, liturgy or some other text, a community of others (pilgrims, the Catholic Church), or laws and moral principles. None of these components is essential or universal, but no religion is without any of them.[9]

When I claim that religion can promote both positive and self-destructive behavior and that both are possible because religion has emerged and persisted owing to its usefulness in early human evolution, to which aspects of religion am I referring? Certainly belief in karma and ritual pilgrimage are too different to fall under the same umbrella! With the great diversity of religious phenomena in mind, I focus on one theme that carries the strongest biological implication: virtually all aspects of religion produce positive feelings that range from sensory delights (feasts and celebrations) to emotional joys, to intellectual and artistic appreciations (reading a scripture, listening to a Mass), to satisfaction in mastering one's weaknesses, and finally, to the most rarefied delights, which are often considered "spiritual." These may include the mystic's rapture, Buddhist nirvana, or even the Dalai Lama's happiness. All of these feelings fall under the broad biological/psychological category of pleasure. I do not contend that pleasure is essential to religion, of course. However, where it occurs, as the previous chapter has shown, our brain is processing information that tells it that things are going well, that adaptation has worked. This, then, is the key to understanding the wide-ranging but subtle biological function of religion.

Religion and Pleasure

So what does in fact cause the pleasurable feelings that religious people experience? It is very hard to say, given the stunningly diverse nature of religion and the equally rich forms of pleasure associated with it. Consider a minute sample of a religious phenomenon that people everywhere appear to enjoy—rituals. According to experts, there are between ten and twenty distinct types of ritual, including rites of passage, marriage, funerals, festivals, pilgrimages, purifications, civil ceremonies, rituals of exchange, sacrifices, worship, magic, healing, interaction, meditation, inversion, and drama. Between them, there are thousands of distinct cultural versions of baptisms, birth rites, puberty and initiation rites, contests and races, calendar rituals, harvests, sowing, political conventions, parades, tree and blossom festivals, new year festivals, passion plays, concerts, royal installations, masquerades, dances, and inaugurations.[10] Rituals draw on colors, sounds, movement, percussion, masking, cross-dressing, eroticism, fireworks, jokes, theater, animals, food, drink, drugs, readings, chanting, burning incense, flinging dirt or dye, and violence. There are also rituals associated with holy wars and crusades, martyrdom, ascetic practices, mystical disciplines, vows, judicial rituals, executions, flag days, memorial days, and military reenactments.

These lists represent a tiny fraction of religious phenomena; they exclude the thousands of scriptures, prayers, myths, and songs that are used as texts either accompanying rituals or justifying them, or perhaps merely coexisting within the same tradition. Some rituals, such as the Jewish Purim festival, are very closely mated to a specific text—in this case, the Scroll of Esther. Others, such as the Carnival in Rio, the Eskimo bladder festival, or the Santa Fe festival, conceal any possible scriptural tie. Some of the rituals play up the existence and majesty of God, Goddess, or gods, while others are merely about nature, a hero, an event, or a place. All constitute religion, which mocks any attempt to limit the definition to a singular spiritual response to God—to faith. In fact, faith may be the least compelling feature of these many kinds of ritual performance, as any honest exchange with the participants may reveal. What stands out first and foremost is the visceral reaction to the events on the street, of which pleasure is very prominent. The pleasure manifests itself in a variety of domains: the sensory domain (food, drink, dance), the emotional domain (the joy of companionship, the loss of inhibitions, optimism), the realm of achievement (completing a vow, mastering a task, discovering new dimensions within oneself), and the cognitive realm (making sense of the liturgy, appreciating justice, understanding a riddle or a joke).[11]

Most important, rituals produce three major types of enjoyment: celebrative, ecstatic, and immersive (which can also be called annihilative). The first, which is the most pervasive, describes a state that most of us have often experienced: joyful participation with others in a festive event. Think of your best friend's wedding, particularly the dance at the reception after the ceremony. Ecstatic enjoyment is probably the experience of participants in a Pentecostal service, those attending a Hasidic wedding service, or dancers and revelers in a carnival—especially after a couple of drinks or perhaps with the aid of other intoxicants. The term "ecstasy" literally means being outside oneself, which is fitting for the feeling of letting go of inhibitions. At the most intense and religiously compelling level of participation stands the immersive level of enjoyment. Here you completely lose yourself. The celebrants report a loss of individuality and identity, a complete merging into something greater. Shamans, Sufi dancers, Sun Dance participants, exorcists, mystics, and even ordinary devotees (*bhaktas*) of Kali during her festival are some of those who can attain this highest level of engagement. It is important to note that the three terms—celebrative, ecstatic, and immersive-annihilative—carry no moral implications. One can celebrate a birth or an execution. One can attain a sense of self-annihilation by merging into God or into a mob. The three terms merely describe psychological states of pleasure as they affect the subject's sense of identity and level of enjoyment.

In the next section, I describe five religious performances in some detail: the Sabbath service, the running of bulls in Pamplona, Spain, the Holi spring festival in India, the mystical disciplines of the Spanish mystic Teresa of Ávila, and the martyrdom of the Sufi mystic Al-Hallaj. These descriptions illustrate the function of pleasure in diverse religious contexts, ranging from celebrative to immersive (or annihilative). I examine the specific source of the pleasure in such religious contexts: is it the object of belief (God, heaven, justice), or is it an invisible current of social forces that runs through all of religious life? It is vitally important to acknowledge this distinction in order to gauge the role of religion in human evolution and to explain the tenacity of religion—or why, in the words of one recent book, "God will not go away."

Four main themes emerge from these illustrative cases:

1. Besides its three levels of intensity (celebrative, ecstatic, and immersive-annihilative), the pleasure of religious performance can be experienced in diverse domains—sensual, emotional, family-based, moral, and spiritual.

2. Religious performances occur within distinct types of social arrangement, from natural families and castes to social organizations based on shared faith alone.

3. The quality of the pleasure in a religious performance is correlated with the nature of the social arrangement. The more natural the group—the family, for example—the greater the emphasis on sensual and emotional pleasures.

4. The true source of religious pleasure is not only or even primarily the object of belief (for example, God), but the underlying social dynamic. This argument, which dates back to the great French sociologist Émile Durkheim, is discussed in more detail in chapter 5.[12]

The Sabbath and Religious Pleasure

THERE IS A STORY in the Babylonian Talmud (Tractate Shabbat 119a) about Rabbi Joshua ben Hananiah. He once received an invitation from Hadrian, the Roman emperor. The emperor asked him about the secrets for preparing the Sabbath meal, which had to be remarkably tasty because it clearly gave Jews such exquisite pleasure on Friday night. Rabbi Joshua taught the emperor's chefs how to make the best foods with the choicest ingredients and spices, but the emperor, despite enjoying the meal they fixed, could not discern anything unusual about the pleasure he received from the food. He summoned Rabbi Joshua again, and the two of them interrogated the chefs. Surely, the emperor insisted, there was some special spice, some rare and secret ingredient or cooking procedure that gave the Sabbath diners their bliss. The rabbi responded that indeed something special did exist, but it was not in the food—it was the love of the Sabbath.[13]

The episode describes two distinct types of pleasure with remarkably different levels of intensity. The first is caused by great food, the second by great food and the love of Sabbath, a distinctly religious feeling. That love is the "hidden spice," the extra ingredient that Hadrian failed to detect. It was natural for Hadrian to confuse the pleasures of dining with the joy of the Sabbath; after all, the food was delicious and beautifully laid out. And so much of the Sabbath takes place precisely around the dining table, with its linen, candles, the singing, and the raising of glasses. Moses Maimonides himself, the greatest individual authority on the Jewish law, said as much. In his influential legal code, Mishneh Torah, Maimonides commented on the following biblical verse (Isaiah 58:13): "And call the Sabbath a delight, and the holy of the Lord clothed with honor." Maimonides wrote: "What is Sabbath's delight? This is explained by the statement of the sages that one should prepare for the Sabbath the richest victuals and the choicest beverages that one can afford" (3.7). If one can afford it, eating a luxurious meal and drinking wine on the Sabbath constitutes Sabbath's delight (3.10). So we wear our best clothes, cover the table with the prettiest white tablecloth, and light the room with Sabbath candles.[14]

But no, the story insists. It's not any of these things, or even the family spending time together and resting from work, that makes the Sabbath a delight. Instead, it's a feeling: the love of Sabbath. What exactly is this feeling? It seems that it is not the same as loving your children or loving a baseball team. Here is how one source describes it. The special joy of that particular day is the obedience to a divine commandment, the response to the highest object of faith, God. Menachem Nahum of Chernobyl, the seventeenth-century Jewish writer, was explicit on this point: "But the truth is that: The joy you should have in fulfilling a commandment is a true spiritual joy, something of the world-to-come. Your joy is in the act itself and in that way do you find joy in God."[15]

Hadrian the naturalist looked for the source of Sabbath pleasure in the recipes for the dishes. We, who are also a bit biased in the direction of feelings, might look to the family and the beauty of the ritual to explain the pleasure. But this may already be a bit too secular. A Protestant, particularly a Baptist, might look to Genesis and its celebration of the seventh day as a day of rest—for the Creator and for those among us who seek to emulate Him. However, this interpretation would not be particularly faithful to the Jewish tradition that promoted the Sabbath. This tradition does not entirely mock the emperor Hadrian, nor does it mock those who explain the Sabbath delight in terms of enjoying the family, the songs, and so forth. The key, however, is obedience to the Law (Torah). It does not matter whether the commandment deals with dietary matters, family affairs, holidays, purities, or any other topic. The uniquely religious pleasure here is the response (assent) to the Law. This response captures something distinctly and irreducibly Jewish, namely, the combination of belief with social action. Obeying the Law is a social act; it places the actor at the center of a sanctified relationship between the Jewish people and God.

The Running of Bulls

In 1926 Ernest Hemingway published his first major work, *The Sun Also Rises*, to decidedly mixed reviews. It featured a cast of shallow characters, aimless survivors of the Great War, who leave Paris for a semiconscious journey to northern Spain in pursuit of a vague notion of happiness. They seek it in Pamplona, where the bulls are set to run at the Fiesta of San Fermin. While Hemingway's book hardly set the literary world on fire, it transformed the sleepy town of Pamplona into a tourist mecca during the days of the festival. Interestingly, the story's most admirable character is Pedro Romero, an accomplished bullfighter, an artist of death, who performs his artistic masterpiece by toying with death in the bullring.

Today, when half a million visitors swamp the old center of Pamplona, the bulls still play the central role in the festival, though not by dying. The images of the *encierros*—the bulls running through the narrow streets splitting the crowds of inebriated men who are dressed in white and red—are as familiar as the old bullfights. To participants in that dangerous rite, the thrill of the run is ecstatic: "The fear goes away and everything goes blank. And when the bulls pass by, you feel extreme relief. You feel exaltation, friendship, life. . . . You're hooked. It's like a drug, and you're almost begging for more."[16]

But the Fiesta of San Fermin is more than the running of the bulls. Beginning on July 6 with rockets going off and lasting until July 14, the fiesta is an enormous and ceaseless musical celebration during which religious themes mix with alcohol, food, dancing, and sexual revelry at a pitch that would bring the complete destruction of any normal society during any other time of year. As one participant, a thirty-one-year-old engineer, explained: "It's not easy doing crazy things when you know that someone you know can see you. But during San Fermin, there's a kind of green light for almost everything. It's like you've put on a mask. You're not yourself anymore but the person you want to be."[17] The fiesta is a typical agricultural festival: a display of material excess when hard workers take time off from producing to consume.

Oddly, at the heart of the huge party, on July 7, is a solemn religious ritual. According to tradition, San Fermin was a convert to Christianity whose teacher, Saturnin, was martyred in Pamplona in 257 CE by being tied to a running bull. Fermin himself was later martyred in France in 303 (by beheading), but his relics were returned to his hometown. The relics of the saint, who may or may not have actually lived, are carried in a procession that travels from the Church of San Lorenzo to the Cathedral of Santa Maria, where a high Mass is held. The ritual is sponsored and carried out by Opus Dei, the conservative Catholic lay movement. The contrast between the sublime and peaceful joy of the Mass and the explosive fiesta pleasures on the streets of the town, between death and life, strikes anyone who is interested in religion; the juxtaposition of a bacchanalian riot to mark the gruesome death of a martyr—an extremely common feature of feasts and festivals all around the world—begs for an explanation.[18]

Saint Augustine, the most influential thinker in Christian history, wrote on the subject in the fifth century and provided the most cogent folk-explanation of the paradox. Commenting on the martyrdom of the two early martyrs Perpetua and Felicitas—both martyred victims of Roman persecution—Augustine explains that the feast celebrating their deaths is not simply a reminder of their courage. The two died on behalf of a permanent and far greater happiness than mere worldly pleasure. Their names, he says, add up to "perpetual felicity": "The only reason, I mean, why all the martyrs toiled bravely for a time by suffering and confessing the faith in the struggle, was in order to enjoy perpetual felicity."[19] The feast, in other words, is not a memorial to fallen victims but a celebration of happiness, the kind of joy one might find only in heaven. Directed at an ancient Mediterranean feast, Augustine's theory would seem to apply to the joys still celebrated on the streets of Pamplona today.

But this was just a theologian's theory. There was always more going on, even in Augustine's own fifth century. The Christian calendar was coordinated with the pre-Christian agrarian festivals, which coincided with the forty-day cycles of the farming seasons. The rural feasts had always been excessive—their point being to put on a display of waste to act as a kind of primordial sacrifice. Later on, Christian pilgrims during the early years of the Church journeyed to the burial places of martyrs and celebrated the dates of local pastoral saints with such worldly passion that in 397 the Council of Carthage tried to put a more decorous limit on the festivities. The pilgrimages, the drinking, the sexuality, the worship of powerful relics—all of these activities looked too much like the fertility and magical rites of the pagan feasts, which the Church had sought to replace.[20] The fact that the pagan festivals that predated Christianity were still celebrated by Christians on the streets of their towns makes Augustine's theory about the martyred Perpetua and Felicitas sound contrived, like a forced and anachronistic etymology.

Nonetheless, Augustine put his finger on something important. Two traditions came together in one festival—first the pagan and then the Christian. As the second, becoming triumphant in the Mediterranean world, repressed the sensual religious pleasures of pagan society, it set aside (tolerated) a day or two per year when the old debauchery could indeed continue. It is no coincidence that such days were matched up with a commemorated martyr, especially one who died in order to reject sensory delights in favor of perpetual happiness in heaven. This is precisely what Christian society—a community of faith—was trying to inculcate among the pagans. And the old pagan violence, to which the martyrs (Perpetua, Felicitas, Saint Saturnin, Saint Fermin) willingly succumbed, now transforms itself into the willingness of young men to risk their lives by running with those same bulls that killed Saturnin. A perfect match is achieved between the two types of religious sensibilities and societies (pagan and Christian) within the festival. The same juxtaposition took a modern turn with the introduction of the *encierros* (running of the bulls) in the late nineteenth century. The bulls were a national and urban symbol parachuted onto a rural and Christian festival that had marked a local identity. The machismo of the national symbol now became integrated into the general bedlam of the festival.[21] Today the violence and the risk of injury, just like the drinking and the sex, as well as the peaceful Mass, are not accidental elements but orchestrated religious and national tools. Most participants today may not know this. But the thrill is acting as though one did, along with throngs of others.

The Holi Festival in the Hindu Village

DECADES AGO, A RESPECTED anthropologist from the University of Chicago, McKim Marriott, published a brief and amusing account of his first experience with Holi—the spring festival celebrated all over India.[22] This episode took place in a small village, Kishan Garhi, across the Yamuna River from Mathura in northern India. This was Krishna country, the place where, according to several Hindu scriptures, Krishna spent his childhood when he was threatened by an uncle who sought his death. The village consisted of twenty-four castes—this was in the late 1950s or early 1960s—including Brahmins at the top and several low castes. The village economy depended on the work of both the lower and higher castes, and the system was based on a careful separation between them (based on laws of purity and pollution). Invited to participate, the American anthropologist witnessed many festive performances—the symbolic burning of a demoness (Holika) in a large bonfire; the tossing of grains, ashes, dirt, or red dye; bawdy displays of sexually suggestive pantomime and joking; the humiliation of several villagers; and the beatings of many others. Marriott was made to drink an unidentified liquid, which he deeply enjoyed. Later he found out that it had been spiked with *bhang* (hashish or marijuana), which explained why he had not been able to report much more of the bacchanalian events that unfolded around him. The next year, still in the village, he recorded the events with greater precision. He discovered that the beatings and humiliations were precisely choreographed: the objects of the violence were those upper-caste individuals (usually men) who during the rest of the year ranked at the top of the rigid social hierarchy. They took their punishment with relish and made little effort to defend themselves. The beaters, in turn, were either boys, lower-caste individuals, or women—

all people at the bottom of the social hierarchy. The beatings were not savage and seemed rather comical.

As an anthropologist, Marriott was aware of the so-called functionalist theories that explained such rituals as ways of reinforcing social values by temporarily reversing them. But he was also interested in what the villagers had to say—that is, in the religious explanation. The participants claimed to be imitating the young Krishna in his village: the boy-god was notoriously erotic and mischievous. The worship of Krishna emulates the theme of love; the beatings with sticks and the abusive taunts reenact the behavior of the outraged cowherd girls who were Krishna's favorite victims.

But there is another theme at play. Two centuries ago, this region between Delhi and Agra was inundated with followers of charismatic devotional leaders from Bengal and from southern India. These leaders, particularly Sri Caitanya, promoted a religion based on a profoundly emotional connection to God (Krishna), such that all other rules of Hindu society (namely dharma) were superseded. These faith-based communities ignored the normal caste hierarchy and accepted members from any level of society to become simple devotees of Krishna and followers of Caitanya. These groups came to this dusty area of India because it was the homeland of the young Krishna. There they instituted the celebration of the god in its symbolic erotic dimension, superimposing it on a preexisting spring festival (associated with the seeding of fields).

The Holi that McKim Marriott observed is two rituals folded into one, but it is also the ritual clash between two types of social arrangements: the caste-based and the faith-based. On the day of the festival all such arrangements are faith-based, and so the violence is not simply about reversing power dynamics by hurting higher-caste members but also about showing and even enjoying acts of self-restraint (by those at the top). Similarly, the debauchery is not about suspending the normal sexual restraints of caste society so much as about demonstrating that the puritanical limits of the faith-based society (of Caitanya) are based on the love of an erotic Krishna.[23]

Saint Teresa of Ávila

Along with Saint Francis of Assisi and Saint John of the Cross, Teresa of Jesus, born Teresa de Cepeda y Ahumada in 1515, is probably the best-known Christian mystic. Teresa came from a wealthy Jewish family in Toledo, Spain, that had been tormented by the Inquisition into converting to Christianity. She had been a rambunctious child and young teenager before joining an Augustinian convent and committing herself to life as a nun. By middle age, following years of intense prayers, vigils, and isolation, she reached a high level of spiritual attainment, with visions and raptures. The most famous image of Teresa is the statue by the Italian sculptor Gian Lorenzo Bernini called *Saint Teresa in Ecstasy*, which depicts the nun in rapture. The Inquisition became interested in the spiritual reports of the aging nun, having looked at her autobiography, *My Life*; they were determined to settle whether her visions were from God or the devil. Her superior, Father Jeronimo Gracian, commanded her to write about her spiritual experiences in the third person in order to throw the Inquisition off her trail. The product was *The Interior Castle*, an extremely vivid and detailed report on the spiritual journey of a mystic.[24]

Mysticism, which is often confused with superstition and fuzzy thinking, is rather a rigorous and direct path to the experience of the divine. It often requires meditation, intense prayers, fasting, vigils (staying awake at night), and a variety of austerities. It has been—and continues to be—practiced throughout the world, from various Christian orders through Hassidic and Kabbalah groups, Muslim Sufis, Yogis, and Zen monks. *The Interior Castle* is an unusually detailed description of both the mystic's techniques and the psychological consequences of the difficult path. Teresa describes her progress in terms of striking metaphors—a castle with concentric rooms, betrothal, and the transformation of a caterpillar into a butterfly. The further into the interior of the castle she moves, the closer she comes to God, the Beloved, whom she ultimately seeks to wed. A light radiates outward from the very interior of the castle, originating from God. The castle is surrounded by poisonous creatures—sins, desires, sensual pleasures—that ambush the seeker of divine knowledge. The exterior rooms, which represent the beginning of practice, are still vulnerable to penetration by such temptations. Beginning with the fourth room and further in, the mystic begins to experience so-called supernatural rewards, including visions, consolations, raptures, and moments of intimacy with the Beloved. Although the journey does not get easier—the closer to God she stands, in fact, the more painful is the remaining separation—the mystic's journey is saturated with joy. This is how Teresa puts it:

> When His Majesty wishes to bless us with supernatural favor, the delight brings with it the greatest peace, quietude, and sweetness to our innermost selves. . . . Anyone who has experienced this knows that ultimately our whole outer selves definitely come to enjoy these sweet pleasures.[25]

What is the difference between ordinary pleasures and divine ones? Teresa speculates about this. She observes that "joyful feelings that come to us from God are similar to natural consolations, except that their source is nobler."[26]

But the most sublime joy is union with the Beloved, a spiritual pleasure so profound that it does not resemble anything in the world, not even other spiritual pleasures. Teresa's description of this experience is undoubtedly her most famous and controversial statement, the one depicted by the Bernini statue in nearly erotic sensuality. Here is how Teresa describes the delicate union between the soul and the Beloved:

> Often when a seeker is distracted and forgets about God, he will awaken her. His gesture is as swift as a shooting star and as resounding as a thunderbolt. Although his call is soundless, the soul is left with no doubt that he is calling her. This is so clear to her that sometimes, especially at the beginning, she may tremble and whimper without any obvious cause for her pain. She feels that she has been wounded in the sweetest way, but she can't figure out how it happened or who inflicted it. All she knows is that the wound is something precious and she doesn't ever want to recover.[27]

Although the journey toward God can bring about moments of extraordinary spiritual delights, the final union transcends even the greatest rapture. Pleasure requires a conscious self who is aware of her own feelings. In final union, the individual identity dissolves into the divine like a stream into the ocean, like a flame into the fire. The two are one—the person who receives joyful rewards from God no longer exists.

Despite the remarkable detail of Teresa's spiritual writings, the subject of pleasure's origin remains vague. Pleasure could be grace from God, but that simply means that Teresa feels undeserving of it (see chapter 7). In fact, God is just as often the source of pain and suffering. There is certainly no correlation between the mystic's proximity to God and her level of enjoyment. And indeed, throughout the mystic's journey, faith in God vacillates in mysterious ways, so much so that doubt is never absent. Some mystics, such as the Sufi woman Rabi'a, have written that the pain and the doubts are more valuable than the joyful moments of divine grace because they challenge the lover of God to offer that which is truly important: selfless service.

Mystics have traditionally been thought of as supreme individualists, people who move against the grain of their own religious traditions, often at the risk of severe punishment. Al-Hallaj, whom we shall scrutinize next, had been a mystic before he became a martyr, and both Teresa and her younger contemporary, Saint John of the Cross, had to evade the eyes of a merciless Inquisition. But most scholars today think of the mystic as the distilled embodiment of his or her tradition, not a spiritual rebel.[28] Teresa's own writings seem to bear this out by being profoundly relational. There is nothing isolated about her external and interior life. She wrote in response to a superior's order, and she wrote for others to read. She wrote about love, and even her God was a loving God, a divine Husband with Whom she sought union in love. Her last reported words, on her deathbed, were: "Beloved, it's time to move on. Well, then, may your will be done. Oh, my Lord and my Spouse, the hour that I have longed for has come. It's time for us to meet one another."

There is something very Jewish about these sentiments, the spiritual eroticism that harks back to the biblical Song of Songs and to the Kabbalistic liturgy of the Sabbath (who is the Bride). In the writings of a Carmelite nun, the eroticism has been entirely distilled, the pleasure matched by pain (both of them equally precious and desirable). What is important to keep in mind is that, even as a mystic, Teresa was never truly alone and her feelings were consistent with the monastic world she occupied.

The Martyrdom of Al-Hallaj

AL-HALLAJ WAS A PERSIAN Sufi mystic who got himself into theological trouble for blasphemy, having been charged with identifying himself with God. He was tried in Baghdad, imprisoned for over a decade, and finally executed in 922. Louis Massignon, the renowned French Islamic scholar (1883–1962), published the enormous, four-volume *The Passion of Al-Hallaj: Mystic and Martyr of Islam*, in which the gruesome execution is described in great detail.[29] The convicted mystic was presented to a large public. Both hands were cut off, as were the feet, and he was crucified (raised and nailed to a gibbet). Finally, he was beheaded. The body was burned to ashes and the head displayed in warning to other potential blasphemers. On the execution itself Massignon quotes numerous sources, including Ibrahim-b-Fatik:

> When they brought Hallaj out to crucify him, and when he saw the gibbet and the nails (*masamir*), he laughed so much that his eyes were weeping. . . . In the second *rak'a*, he recited the *Fatiha* and the verse (Qur'an 3:182): "Every soul shall taste of death, and you shall be paid your wages on the Day of Resurrection; whosoever is saved from Hell and admitted to Paradise shall be blissfully happy: for what is the present life but the precarious enjoyment of vanities?" . . . I beseech You, O my Master, to grant me the grace to thank You for this joy that You have given me, to the point of concealing from everyone else's sight what You have revealed to me of the flames of Your countenance (which has no form), and of forbidding to any other the

> look that You have permitted me to cast on the hidden things of Your (personal) mystery.[30]

There were other mystics in Baghdad and other preachers, but Al-Hallaj was unique. After a long period of spiritual introspection in Mecca, he assumed for himself the power to speak of God in the first person, as if the divine essence and his own were merged. He preached in Baghdad to anyone who would listen, including low classes and women, though few could understand the full extent of his spiritual insights. He was a charismatic leader, a savior, who was said to perform miracles and inspired his followers to intense devotion. All of this, but particularly the first-person witnessing on behalf of God, represented both a theological and social threat to the entrenched religious authorities, the *ulama*. It completely revolutionized the customary role of the religious teacher and usurped the foundation of existing religious society—basing it on the charisma of an individual. It was thus not only inevitable that Al-Hallaj would be tried and sentenced but also that his fate would happily confirm his own social and religious assumptions.

The dramatic example of Al-Hallaj's martyrdom highlights the same important questions that run through the other four cases:

- If pleasure has evolved from adaptive success, how can impending death be accompanied by such a display of pleasure?
- If pleasure can be felt *despite* imminent execution, what causes the pleasure: the faith in heaven or the special knowledge to which only the mystic has access?
- Are faith in heaven and special knowledge religious? Are they adaptive?
- Are faith and special knowledge the true causes or mere objects of Al-Hallaj's pleasure?

The other cases are a bit more mute but similar: What precisely is the pleasure of obeying sacred law (Sabbath)? What is religious about festive joy, or about beating up your superiors in the name of Krishna? And what, finally, is "rapture"? All of these situations bear a resemblance to the earlier example of the London immigrant who wears the Islamic *niqab*. Is it possible that, in her case as in Al-Hallaj's, the pleasure of the act is caused by some nonconscious factor and she falsely attributes her pleasure to an object that comes to mind as she experiences the glow of pleasure?

But how could one possibly confuse the object of pleasure—its apparent cause—with the true cause? Peter Kramer (*Listening to Prozac*) discusses this phenomenon with reference to drugs that make us feel good ("mood brighteners") in separation from the true objects within the world (such as altruistic actions, prudent plans, intellectual achievements) that ought to be the true causes of our good feelings.[31] This is the issue of "separability" (I have called it the "Prozac effect") that Kramer thinks can lead to hedonism or addiction. We feel good for chemical reasons and project the feeling onto false causes (perceived objects). We all experience this effect after eating chocolate or drinking strong coffee. The "high" we get from these substances may make us feel good about ourselves, make a commute tolerable, or even help us laugh at a bad joke. The reason it is so easy to be wrong about the true causes of pleasure is that we are unaware of what takes place in our brains, which is where pleasure is born.

If the distinction between cause and object is valid, it is possible that Sabbath celebrants, festival revelers, mystics, and martyrs receive pleasure from some cause that is not religious at all. For example, Al-Hallaj may have been psychotic or paranoid, or perhaps he experienced delusions of grandeur. One might also argue that he believed in heaven and hell and, with a clear and sane mind, enjoyed the thought of his enemies roasting eternally in hell (as a holy *schadenfreude*). It is also possible that the presence of his many followers, his very public death, thrilled him with the thought of eternal fame. All of these possibilities would imply, rather cynically, that there was nothing particularly religious about the martyr's joy. Similar considerations might apply to the other cases. More likely, however, the martyr was acting on behalf of that small community of followers, the few who were initiated to true knowledge of God. His pleasure, which was indeed thoroughly religious, relates to this aspect of his life, namely, the social. It is the social context that makes both the religious and the pleasurable possible.

Religious Pleasure and Adaptation

Religious performances are highly textured and vary enormously. No general theory of religion can ignore this or gloss over the particular and odd facts of local practices. Still, is there any discernible pattern among these five cases, and can we learn anything of general value about religion and pleasure? The answer is yes, but before getting to it, consider the following two scenarios.

Two groups are taken to New York to watch a play on Broadway. The first group pays the full ticket price for a specific play; there is no option of seeing any other. The second group receives free vouchers, which they can use at any of five plays. At the end of the night, members of both groups are asked to rate how much they enjoyed their play. Which group would rate the play they saw more highly?

In the second scenario, two groups join an exclusive club: one group must pay a great deal of money and also pass an initiation test, the second group receives free admission. Both groups are then asked to rate a number of club activities in terms of enjoyment: which would rate them more highly?

If you answered by saying that the payers would rate their pleasure more highly, you are right. But why? There are two types of answers to this question, both in their own way related to evolutionary psychology and hedonics (pleasure theory). The first we have already encountered with Mrs. Keech at the beginning of this chapter. The *cognitive dissonance* of paying more or suffering more to obtain something will compel individuals to rationalize their expense by reporting (to themselves!) greater satisfaction. It is important to note that cognitive dissonance is a source of profound discomfort, so the pleasure reported may actually be the removal of a jarring dissonance.

Second, according to *choice and prospect theories*, the high cost of having obtained something desirable reduces the attractiveness of the optional objects that were excluded and increases the investment in that one object or event. In other words, attaining that one thing imposes a restraint and minimizes our freedom. The more invested we become in one thing, the more we lose the opportunity to opt out or to choose something else. Often cited today is the case of sticking with an ill-considered war whose costs keep going up. U.S. military forces are still in Iraq because we have already invested so much there. Oddly, such facts also result in reports of higher satisfaction in the original choice. For example, people who choose to marry instead of living together report higher levels of happiness, despite the fact that their choice in sexual partners disappears. Paying more for a television in a high-end electronics store than at Wal-Mart often leads to greater satisfaction in the brand than comes from purchasing a cheaper one.

The general principle appears to be that the greater the cost of a decision, the greater the reported benefit (pleasure). Participants in such experiments rarely understand why they report their particular level of happiness—they may think that they are simply responding to the quality of an item, an exclusive club activity, or a marital partner. In fact, the cost of their choice and the removal of options have a great deal to do with their level of satisfaction. All of this may be invisible to the person who makes the choice. According to basic prospect theory, the fear of loss exceeds the advantage of gain.[32] If you were offered a 50 percent chance at gaining $200 or a sure $100, which would you choose? The vast majority of people opt for the sure money. But when choices are limited, fear of making the choice that will lead to loss (choosing the $200 or losing the coin toss) is diminished. At such times, surprisingly, raising the cost of the choice actually diminishes the discomfort of not having chosen the few other options. People seem happier with fewer choices and more content when they make an expensive one. That is where prospect theory and cognitive dissonance combine.

How does all this apply to the five rich examples from the world of religion? Some religious groups are completely organic and involve no choice at all. This would include a Jewish family, a Hindu caste, or a tribal religion. Others are based strictly on choice: the monks or nuns who have opted for monastic life, or the followers of a heretic or charismatic leader. These can be called communities of faith—or in Durkheim's similar term, "moral communities."[33] There is one general pattern with regard to the two types of religious groups: the more voluntary the nature of the group, the greater must be the cost of joining. This often entails giving up wealth, comforts, or health.[34] Many renounce their families and undergo severe physical and emotional tests. This pattern is not always easy to discern in the five examples owing to historical and sociological factors. The participants in the Holi festival, for example, are drawn from both elements (caste members and followers of a devotional movement). The visitors in Pamplona include tourists who are there just to have a great time. But the case of the Sabbath meal (family) is clear, as are the cases of both Teresa (monastic group) and Al-Hallaj (heretic sect), so we shall focus on these. The joys of the Sabbath are not surprising because there is no downside. What is surprising is the report of pleasure (rapturous delights) by Teresa and Al-Hallaj, in whose situations the costs were so high. So what is going on? In the case of the Sabbath, there is no choice of membership in the family that celebrates (and no regret over not having made a bad choice), while a nun or a follower of Al-Hallaj (or he himself) could have made a different choice. The high level of joy reported by members of the voluntary groups is related to the level of their costs—the hardships and the suffering inflicted on them. The high cost makes up for not having made some other choice and regretting that. Thus, participants in the Sabbath service and mystics as well as martyrs report feelings of satisfaction, but for different reasons within the same general rule.

Here we need to add another factor. The pleasures of nonvoluntary groups with low costs of participation tend to be sensory and emotional. Meanwhile, the pleasure associated with choice-based groups for which participation costs are high is usually described as nonsensual: it is experienced as spiritual or intellectual and relies on terms such as "happiness" and "spiritual consolation" or "rapture." To return to the discussion about adaptation and pleasure that closed the previous chapter and will continue in the next chapter on happiness—the pleasure of the first group (the family) is replenishment or novelty pleasure, while the pleasure of the second group is mastery pleasure.

I am aware of oversimplifying a very complex picture in making broad judgments about entire traditions. For instance, enjoyment of the Sabbath meal can be profoundly mystical, such as when it is celebrated not by the family but by a Hasidic group known as *havurah*. Membership in such a group is based on choice and on the sacrifice of more basic pleasures. It is in such contexts that we can explain the ecstatic foundation of the Sabbath meal as obedience to the Law as opposed to something as mundane as rest or family togetherness.

Note that I have said nothing about the nature of God, the contents of teachings, beliefs in the afterlife, and so forth.[35] The principles that regulate the experience of pleasure (how much increased pleasure you get from an act or a choice) are independent of those religious themes, however important they may be in their own way. And furthermore, the members of religious groups always attribute their decisions to a god, a teaching, a scripture, and so forth—not to the social organization, and certainly not to pleasure itself, which is an invisible psychological force. This "invisibility" of pleasure is precisely what experiments have shown. Here is one example.

Michel Cabanac is a physiologist at Laval University in Quebec. He asked students to rate the pleasure they got from playing a video game. The students then sat in a temperature-controlled room and Cabanac, while gradually cooling it down, asked them to rate how unpleasant the feeling was. He then combined the two experiments. In an interview with *The Guardian* on December 16, 2004, Cabanac noted: "We cooled the room down, and every time, the same thing happened. As soon as it was cold enough for their displeasure rating to just outweigh the pleasure of playing the game, they stopped the experiment." Cabanac concluded that the tests show that, regardless of what we think about how we make decisions, our choices are ultimately driven by pleasure. "Pleasure is the common currency that allows us to make any, and I mean any, decision in our lives," he said. "Any decision is made according to the trend to maximize pleasure." The reasoning we think is motivating us—our conscious explanation—is folk theory; it is not entirely wrong, but it describes only that aspect of a phenomenon that takes place at the level of conscious awareness.[36]

The general rule, to summarize, is that the greater the costs of involvement in voluntary religious groups, the greater the level of happiness and at the same time the more advanced (from an adaptive point of view) is that happiness. Religious pleasure, to put it briefly, is the reward for the high costs of the social choices we make. One of the principal functions of religion is precisely to allow us to remain social in increasingly complex ways and to enjoy doing so through happiness. Religion acts in an adaptive way by transforming simple replenishment pleasures (those of the senses) into the more advanced pleasures of the mind. Only by means of such an education could a Teresa or Al-Hallaj enjoy their experiences despite the enormous costs. How this is done is the subject of the next chapter.

CHAPTER 4

The School for Happiness

We have now come to the part of the book that people may find hardest to accept, namely, that happiness is a form of pleasure. Most of us think of "happiness," in the standard dictionary definition, as a state of well-being or contentment, and it seems to us unrelated to the more lowly sensory experience of pleasure. In common usage, pleasure is something we feel within the body, while happiness seems to belong in the mind.[1] Most of us would say that happiness depends on meeting certain values and goals, such as paying off our mortgage or putting our kids through college, whereas pleasure just "feels good." In this chapter I show, however, that happiness is a complex product of pleasure. To use a term from the sciences of complexity, it is "emergent."

Emergence is the gradual building of something complicated that exhibits a great deal of intelligence in its design and in the way it meets its survival goals, like an ant colony or a city neighborhood, from the simplest elements to the most complex without an overall plan. However, the complex final product is not to be confused with the sum total of its elements. Some of the most successful urban neighborhoods, and certainly ant colonies, are not planned in advance with their final design in mind. Steven Johnson, who wrote an accessible book on this difficult subject, illustrates emergence in the following way.[2] Imagine a pool table where the balls are in constant and random motion propelled by small motors. They collide, reverse direction, spin, and create a general frenzy on the tabletop. The rules that guide their behavior are simple physical and mechanical principles, and it is highly unlikely that any discernible order will emerge—for example, that all the even-numbered balls will end up on one side and the odd-numbered ones on the other. So what would it take to generate some recognizable pattern? Imagine that the balls are programmed or designed in such a way that they swerve left after colliding with a solid-colored ball, accelerate after contact with the three ball, stop dead after hitting the eight ball, and so forth. There is no overarching rule telling the balls to fall into a specific pattern, only local rules that are internal to each mindless "agent" (ball) as it moves on the surface. These rules can, in time, produce a complex pattern, such as forming a triangle of billiard balls.

But this is not enough, because a triangle serves no purpose and there is no reason why the balls should retain that shape or repeat it. As Richard Dawkins demonstrates so vividly in *The Blind Watchmaker*, if you are going to have a very large number of monkeys with typewriters produce Shakespeare plays over a long period of time, you need to have some ratcheting mechanism that keeps the correct letter combinations in place as new ones accumulate toward the final product.[3] This mechanism is not a grand design concocted by a Grand Designer, a heavenly Shakespeare. Instead, adaptation does that work, and the purpose is survival and reproduction. Those patterns that serve the survival agenda of the gene tend to persist because the genetic rule that produced them gets passed on.

What is truly important in the evolutionary theories of complexity is that the emergent product owes its existence to simpler, mindless elements along with a small number of intrinsic rules that account for the elements interacting in a bottom-up trajectory. Happiness, I believe, is such a product relative to pleasure.[4] Recall that pleasure results from simple adaptive responses to situations in which the body has gone out of equilibrium—for example, becoming dehydrated. Pleasure is the reward for the proper solution to the problem. The brain's feedback systems light up in the appropriate places, resulting (eventually) in the awareness of discomfort, perceived as thirst. Drinking is a pleasure-seeking act, but the pleasure is the reward of hydrating depleted cells—a correct biological decision.[5] Such behavioral adaptations look a bit like a cycle or a wave, with three main parts: *novelty*, *response*, and *restoration* of the original state, followed by novelty again, and so forth. I show that there are a small number of simple rules that apply to this cycle and that when they are properly executed pleasure becomes happiness. In this chapter, I show what these rules are and how they work. To give one key example: a simple time delay between novelty and response is essential if happiness is to emerge. A person who learns to delay satisfying his need to drink in response to the first pangs of thirst begins to perceive thirst differently, both as a body feeling and as a problem that needs to be solved.[6] A feeling of mastery over the sensations of thirst and over the rush to drink produces a different type of pleasure—achievement or mastery pleasure. This is one essential condition for obtaining happiness, and it can be learned.

But why should all of this matter? What difference does it make whether happiness emerges from pleasure or whether it is something else altogether—say, a cultural value, like haute couture relative to clothes? We saw in the previous chapter that religion delivers both pleasure and happiness. There are both feasts and celebrations (pleasure) as well as what Saint Augustine called deep and lasting happiness. This book is largely about martyrs, mystics, and religious warriors who wish to suffer and die for some vision of supreme happiness—for their God. God Himself is often identified with happiness—either as the source or the very essence of this feeling.

If you were to tell religious people that their religious happiness, which they often experience as grace or love, is an evolutionary product, you would probably meet with ridicule. I am not saying that "God" (whatever that means) is a product of evolutionary process—not in this book at any rate. I am saying that the human response, the feeling, is such a product.[7] That is precisely what makes religious actions so appealing and intractable. But there is another point. I showed in chapter 2 that people often attribute their pleasure to an external cause with no awareness of what their own body contributes to the experience. We can call this the "hedonic attribution error," or if that is too much jargon, the "Prozac effect."[8] The subjective feeling, the mood, has distinct neurochemical causes, but the patient comes to feel that the world itself is more benign. The same is true for happiness: we are aware of the object—the attributed cause—of our happiness, but remain unaware of the mechanism that brings the feeling about. Take, for instance, a group of young people who join the Marines. They may operate under the impression that they are acting on, and are satisfied by, the compelling power of an argument (freedom is the basis of democracy), a cherished idea (*semper fi*), or a persuasive value (patriotism), while the decisive motive for their action is a wish to be happy—or, to be blunt, to experience pleasure. As opposed to a conscious sense of duty, such a misperception is held by some but hardly all recruits. In contrast, the actions of martyrs, mystics, and suicide bombers are both profoundly hedonic and misperceived. It is therefore vitally important to understand the mechanism that produces all states of happiness.

I use four examples to demonstrate how pleasure becomes happiness: three of them are fundamental (food, stories, music), and the fourth is religious. The first three show the operation of the simple emergence rules (and how children learn them), and the fourth illustrates their application in the religious training for happiness.

Learning to Like Food

AT ITS MOST BASIC level, food pleasure is adaptive in that it restores the body's balance when it becomes depleted, dehydrated, or lacking in calories, minerals, salts, fats, and so forth. In other words, it is essential for survival. You do not need an advertising agency to tell you how much children love apple juice, with its sweet flavor and easy calories. It soothes the most miserable toddler and fixes every injury. And honey-roasted peanuts, as mentioned earlier, are the perfect evolutionary snack: sweet, salty, and fatty all at once, they contain all three of the basic ingredients for replenishing a body. I would venture a guess that if our ancient forefathers had found this snack lying around, they never would have invented tools, hunted, or developed societies. Finally, every ballpark in America is a laboratory of evolutionary snacking: first the overly salty popcorn or peanuts, then the too-sweet soda, then the hot dogs—one after the other, like metabolism itself.[9]

But how do you get from the basic food pleasure, from apple juice and peanuts to something more elevated and complex, like an expensive Clos de Vougeot? Recall Isak Dinesen's wonderfully ironic short story, "Babette's Feast," with its room full of austere Christians forced to enjoy the superb cooking of Babette, who had been a great Parisian chef.[10] One of the diners, Colonel Lorens Lowenhielm, a widely traveled and sophisticated man, has no such difficulties. His eating rapture, in the film version, is our cue to appreciate just how great the meal of tortoise soup, game hens, and so forth, truly is. Meanwhile, the Christians provide the comedy because they love the food against their will (and conscious awareness) and in the face of their most cherished world-denying values. The pleasure dissolves their petty disputes, jealousies, fears, and self-righteousness.

So how do we become sophisticated eaters? Kids hate champagne, espresso, goat cheese, and similarly "complicated" foods and drinks. How do we teach children to enjoy hot spicy chili? By using the same mechanism that turns simple pleasure into educated taste and that is also, as I show later, the root of happiness.

This is an evergreen topic: you can look up back issues of *Good Housekeeping*, *Good Parenting*, or even *Parade* magazines for expert advice on how to get Joey and Jenny to eat their vegetables. These six basic techniques are not all equally effective, and not all of them are about getting the kids to actually enjoy healthy food. I explain and illustrate each in turn and then discuss the basic rules behind the good ones.

1. *Behavior modification (punishment and reward):* Classic behaviorists do not think that pleasure, as we commonly understand the term, actually exists. The goal is simply to effect new behaviors with either punishment ("negative reinforcement") or reward. If you have ever had food punishment inflicted on you or tried it on your own kids, you know that it is mostly useless. You can make your child stay at the table until the liver is all

gone—as I recall from my own childhood—but you will not get the child to enjoy it. On the contrary, liver will probably become a lifelong aversion. Rewards can be more effective, when cleverly executed. You can offer inducements, such as dessert or a reduction in the number of bites of spinach the child has to take. Sometimes social rewards may be effective, such as praising the toddler who tries something new or finishes her carrots. But social rewards come with some risk if the inducement is not directly related to the eating achievement and appears—to the child—to tie his performance to his parents' love or acceptance.

2. *Taste masking:* I call this the Starbucks or Salad Bar Syndrome. How do you get kids to enjoy a bitter drink like coffee or to eat vegetable salad? You add milk and oversweeten the drink (think of a tall hazelnut latte) and you drown the vegetables in ranch dressing. A first-time visitor to an American restaurant would have to conclude that Americans detest lettuce, cucumbers, and tomatoes and that plain coffee is something that only detectives on television drink, while making a sour face.

3. *Imitation:* Imitation is far more effective than reward when done early, persuasively, and sparingly. Facing your child on her high chair, you dip the spoon in the jar of creamed squash and bring a spoonful to your own mouth. If she calls your bluff, then you put the spoon in your mouth and swallow. Then you smack your lips and make rapturous sounds, the kind Andy Griffith used to make on his Ritz crackers commercial.

4. *Easy access:* A recent issue of *Parade* magazine suggests making peeled carrots easily available and making sure potato chips

are nowhere to be found. This sounds like weight-loss advice for adults, but the technique basically works. If your kids do not get a hold of chips at school, they probably will learn to enjoy the crunchiness of carrots.

5. *Benign masochism:* This is a daunting technical term for a method uncovered by food psychologists.[11] You could call it "pretending is fun," and it works even for painful behaviors like eating hot chili. It works like this: Think for a moment about riders on a scary roller coaster or an audience at a horror movie (probably the same people). What causes their thrill? It is precisely the various sensations of falling, losing control, extreme speed, perhaps even struggling with nausea. Toss these same people out of an airplane without a parachute or put them in a runaway train, and these very same feelings would be truly horrifying. But knowing that they are safe (personally, I hate roller coasters and horror flicks), these paying customers enjoy the thrill. According to several psychologists of food, especially Paul Rozin of the University of Pennsylvania, the same principle accounts for the way children (young teenagers) come to enjoy spicy foods, and it may apply to frightening foods (sushi) as well. A child learns to enjoy the burn of the chili because it is a safe pain, and experiencing it demonstrates mastery over childish fear. Benign masochism demonstrates a mastery, through the use of reason or due to social motives, over emotions such as fear.

6. *Opponent process:* This term sounds even worse than the previous one; not surprisingly, the two are related. The behaviorist R. L. Solomon discovered this mechanism in 1980, and what he described is a perfect illustration of the possibilities hidden within biological adaptation.[12] Unlike benign masochism, in

which thinking tells the chili eater that it is safe to burn, opponent process describes the way the eater enjoys, not the painful or frightening features of the food, but the way the body responds to it—or any other source of threatening stimulation. In the immortal words of the *Kama Sutra*: "Slaps and sighs are also part of sexual practice." Think of a dog in the bed of a pickup truck as it rounds a corner. The g-force seems to push against the dog, and he has to shift his weight. The first force is process (a), and the response is process (b), the opponent process. As we respond to new stimulations (thirst, noise, sour taste), the body generates various responses aimed at restoring the body's original balance. A sour apple or hot chili makes your mouth salivate in order to cool down. Solomon suggested that we can train ourselves to enjoy the response (b) and even magnify it. With spicy food, the pleasure comes not from the burn but from magnifying the body's response to the burn, namely, the cooling-down process. (It took me a bit of time to learn, while living in India, how to enjoy hot tea during extreme heat waves.) Usually, the cooling effect is achieved with a cold beer or a soda that makes the spicy chili more enjoyable. (Note that the chili also makes the bitter beer tastier—potato chips, chili, pizza, and similar richly spiced food make it possible for a new drinker to enjoy beer. This effect, however, is not opponent process but taste masking.) To give a different example: Skiing gives distinct types of pleasure. One is the extended falling, which feels good because the body's response to falling is adrenaline, heavy breathing, and rapid pulse—the "rush." The opponent process is the mastery over the skills that keep the body upright and under control in response to gravity and speed.

What is the general lesson from these basic methods of refining food tastes? Remember, we are looking for simple, "mindless" rules—algorithms, really—that turn simple phenomena into complex ones. The simplest rule here is temporal. The child needs to delay her reaction to something undesirable, frightening, or distasteful. Delay implies, by definition, a basic awareness of a gap between the new stimulation, the feelings it evokes, and the response. This space opens up opportunities for evaluation, imagination, and planning, for becoming aware of oneself as an actor, and so forth.[13] Obviously, there is a huge difference between a two-year-old who learns to like carrots, a twelve-year-old who learns to like spicy chili, and a connoisseur of fine wines. Only the older child learns to capitalize on benign masochism by thinking about the burn. A toddler, or even a six-year-old, cannot do that. And only the thirty-year-old (or so) can discover opponent processes in the subtle qualities of some regional wine when looking for "a combination and succession of sensational elements—from the colour of the wine, through its smell, to its taste in the mouth."[14] But at bottom, all of these individuals are doing the same thing: slowing down their initial response to novelty and inserting new elements into the adaptive process (thinking about the context, reflecting on the response while ignoring the stimulus, figuring out the puzzles of flavor, and so forth). As these skills develop, the eater and the drinker come to experience dining as highly satisfying and as far more than just a way to eliminate hunger.

Literature: "The Three Little Pigs"

THE READER MAY BE forgiven for feeling a bit lost. I have yet to talk about religion—let alone religious violence—in this chapter. Where is the thread? At this point what I am saying is that learning to enjoy religious beliefs and actions is an acquired taste, like learning to eat your spinach. It does not happen naturally—a child will not learn to enjoy prayer, or for that matter read the four gospels, unless she is taught to do so. On the other hand, religion is truly nothing if it is not enjoyed in one form or another. After looking at how food habits set the stage for learning new pleasures, we can now move on to analyzing a more abstract skill—enjoying literature.

"The Three Little Pigs" is a nursery tale/fable that never seems to lose its appeal. Dating back over half a millennium at least (wolves had largely disappeared from Europe by 1500), it has seen countless published versions, including, among the most recent, an MTV clay animation with Green Jelly performing the Rock soundtrack. Because the narrative is a fable rather than a fairy tale, the message is crystal clear and highly moralistic: plan ahead and work hard or face doom. This clarity survives the many competing versions despite drastically changed detail. In the first published version (by James Oscar Halliwell), only the third pig survives all the devouring that takes place, including that of the wolf. The three houses are built of mud, cabbage leaves, and brick (no straw or wood). The first two pigs are fond of wallowing in mud and eating cabbage, respectively. Naturally, that is the preferred version of Freudian theorists, such as Bruno Bettelheim in his landmark discussion of Grimm's fairy tales, *The Uses of Enchantment*.[15]

The most familiar version of the fable, based on the publication of Joseph Jacobs (1898), begins with four rather musical lines:

Once upon a time when pigs spoke rhyme
And monkeys chewed tobacco
And hens took snuff to make them tough,
And ducks went quack, quack, quack, O!

The first pig builds a house of straw. A wolf comes knocking on its door: "Little pig, little pig, let me come in." The pig replies: "No, no, by the hair of my chinny chin chin." And the wolf says: "Then I'll huff, and I'll puff and I'll blow your house in." The entire affair, properly read by a willing parent, is nothing less than a musical performance, a mini-opera. In this version, the first two pigs survive the destruction of their homes and take shelter in the brick house of the third pig. The second part of the story has the pigs taking the initiative and fooling the wolf twice (in a turnip field and an apple tree) and finally landing him in a boiling cauldron as he scampers after them down the chimney. They eat the wolf and live happily ever after.

There are obvious didactic reasons to tell this nursery tale to your kids, especially if you are a Victorian parent and, a Marxist critic would remind you, a bourgeois brainwasher of young minds. After all, you want to raise your children to be hardworking and alert, perhaps even a bit crafty. But why sing it? And why do your children ask to hear it over and over again? Is it your amusing performance, or something about the story itself?

For Bruno Bettelheim, the very subject of the story is pleasure. Moral lesson and narrative form conspire to create a seamless learning and growing-up experience. The narrative (Bettelheim incorrectly regards it as a fairy tale) tells the story of the child's renunciation of the pleasure principle in favor of the reality principle. These are two Freudian concepts that apply to the infantile wishful fantasies that center on (mostly oral) pleasure, followed by a growing sense of reality. Bettelheim writes: "Living in accordance with the pleasure principle, the younger pigs seek immediate gratification, without a thought for the future and the dangers of reality."[16] The story entertains the "nursery-age" child because it teaches in an enjoyable manner that one must not be lazy and must recognize the dangers of the world outside the id-dominated personality. The enjoyment comes from subserving fantasy to knowledge, or as Marina Warner puts it, allowing the "dreaming" to represent the "practical dimension" of life to the child's vivid imagination. In other words, this story, like all fables, helps the child grow out of her infantile stage of life and into the next one, which requires a bit of realism along with some restraint.[17]

But why do children enjoy fantasy, and can the fantasy remain formless? Try telling the story of the three pigs with the third pig first; have him dine on the wolf and narrate the story of the wolf and the pig's two younger brothers, perhaps as he washes his soup bowl after finishing the stew. Adults may appreciate the change in narrative structure, but your four-year-old will not get it. "Daddy, tell it the right way!" she will demand.

According to Rony Oren, who heads the Department of Animation at the Bezalel Art Academy in Jerusalem, every student who wishes to become an animator must learn the basics of script writing for animation shorts. The central elements of the basic script, according to the syllabus, are the initial exposition, the turning point, conflict, resolution, and afterward. Students discover that this pattern holds in all of the dozens of film scripts, cartoons, and feature-length animation movies they analyze. All of the known versions of "The Three Little Pigs" abide by it as well: the exposition is the description of the pigs and their mother; the turning point is the mother sending the three into the world; the conflict appears in the person of the wolf; and so forth. If you begin your film with the three pigs eating wolf stew, you have misplaced the conflict and lost the drama that it generates for young viewers. In fact, your version is likely to become a farce. You have sacrificed the exciting and suspenseful elements of the plot, which, according to the renowned writing teacher Laurence Perrine, are essential elements of any enjoyable story. Without the drama, the triumph of the hero is taken for granted and the happy outcome loses its impact on the listener or reader.[18]

All of this well-known hedonic information applies to the young child (consumers of cartoons) or even to those who make do with the gratifications of pulp fiction or prime-time television shows. This effective, if clichéd, formula is found in many types of narratives around the world and throughout history, and no mastermind scripting is required to spread so sensible an entertainment formula. One need not even read Aristotle, who wrote (in *Poetics*) the first scientific analysis of drama. Good (enjoyable) scripts, simply stated, are adaptive.

The three basic parts of the script (exposition, complication, denouement), according to this idea, correspond to the three primary stages of adaptation in evolutionary psychology: prior state, change, and restoration. Recall the thirsty athlete in chapter 2, training in August for the new season. His body feels the loss of liquids as thirst, which emerges in his mind as an unpleasant, perhaps even painful, disruption. In this mini-drama, drinking is a response, obviously, a pleasant resolution to the moment of crisis (if I may be forgiven the overly dramatic terms). The story of the three little pigs is also unmistakably about survival—what else is the killing of the wolf?—but that is not its most interesting point. In fact, it is the pleasure in the hearing of the story in the correct manner—the literary performance—that points to evolutionary survival, not the content of the story. The story could be about winning the hand of a princess in a contest without losing one's head, or about gaining the throne against two older brothers. The evolutionary-psychological aspect of the script rests in the exposition, and the trace it leaves is in the form of the enjoyment we get from hearing or reading the script in proper sequence—regardless of the subject. The "proper sequence" is a structural term, like the proper sequence of notes in a pleasing tune. This structure is very basic, whether you're reading the Bible, Shakespeare, or a fairy tale.[19] Adaptation and enjoyment are interlinked in the structure or sequence of the narrative. But it is vital to note that they are also mediated, at a slightly less elementary level, by means of emotions.

Emotional adaptation, at its own most basic level, responds to the successful "script" by means of arousal, turmoil, and relief. These can manifest more concretely as curiosity, surprise, disappointment, fear, frustration, anger, suspense, hope, relief, and joy. If you eliminate the suspense or the fear (by changing the script sequence), then relief and joy also dissolve. But the script does change, eventually, as literary works become more subtle and as children grow up to be more sophisticated readers. While literary theorists discuss what makes literary works good (and pleasure does not rank very highly these days), the hedonic psychologists ask what it is that enables the naive reader to become a true connoisseur as his enjoyment moves from nursery tales to great works of literature.[20] How does taste become educated, or how does pleasure grow more complex? Note that there are two distinct but interrelated questions here: How does the reader become more sophisticated in her taste? And how do works of literature develop ever-more-elaborate devices ("scripts") for deepening and enriching the reader's pleasurable experience?

These types of questions are more familiar to psychologists than to art critics. The qualities that define a great work of art may not be those that best satisfy hedonic responses. A work of shattering political impact—a photograph of an execution, a biography from the death camps, or even a portrait of the right kind of person—may leave the viewer or reader with a sense of suffering, unmitigated by the aesthetic effectiveness of the medium.[21] Our purpose here is to study the education of taste, leading to the point where we recognize that religion, more so than even art, is hedonic in its very core.

To summarize the story of "The Three Little Pigs," the child's enjoyment of the fable depends on effective scripting in such a way that the emotional responses are truly evocative. The sequence of events in the story, the syntax or plotting of the events, and the sequence of emotions are interrelated. Of course, even a simple story evokes different types of pleasure because several levels of script are playing at the same time (like the counterpoint of a tune). For example, if the story is illustrated, the child will respond to the visual images as she hears the story. The vocal manipulations of Daddy's voice can produce another script (perhaps silly, undermining some of the fear). Furthermore, the young child insists on hearing the story repeatedly. In fact, "the driving force for a preschooler is not a search for novelty, like it is for older kids, it's a search for understanding and predictability."[22] So how does the simple (central) script become elaborated, and how does the child learn to enjoy the early elaborations? Is it anything at all like turning a simple little tune into a slightly more complex one?

Music: "Twinkle, Twinkle, Little Star"

THE OLD FRENCH TUNE known as "Ah! vous dirai-je, maman" is just as familiar as "The Three Little Pigs." American kids know it as "Twinkle, Twinkle, Little Star," "Baa, Baa, Black Sheep," and also the Alphabet Song.

Twinkle, twinkle, little star,
How I wonder what you are!
Up above the world so high,
Like a diamond in the sky.
Twinkle, twinkle, little star,
How I wonder what you are!

Mozart composed a set of twelve variations on the tune (K. 265), which, though currently available in a variety of recordings, are entirely unfamiliar to virtually all the children who know the basic tune. The composition begins with the theme that we all know. It is a highly symmetrical melody that opens with an ascending phrase that consists of three sets of repeating notes, followed by the single seventh. The phrase feels a bit like a question and leads to the corresponding answer, a descending phrase of the same structure. The following phrase, which repeats, begins with a surprisingly high note (on "Up") and then moves down the scale and repeats itself. The melody and lyrics then revisit the opening phrases.

For a three-year-old, there is a lot to like in "Twinkle Twinkle," in terms of mastering both the words (the simile) and the melody. The tune articulates a question, an answer, a striking declaration, an explanation, and, finally, closure. Interest is aroused, tension builds up, an evocation takes center stage, and then tension dissipates as interest is satisfied. All this takes place in the simple folk melody.

Mozart stays very faithful to the original, although he adds a few minor ornamentations (for example, under "are" in the second phrase and "sky" in the fourth). But even his initial statement of the basic theme is extremely sophisticated, owing to counterpoint. The melody is played in the upper voice, with the right hand of the pianist, while a second, harmonizing, lower vocal tune matches it, note for note. The counterpoint melody does not sound like the theme at all, but the cumulative effect enriches the clearly recognized melody and gives it a pleasing harmonic depth. Even a child would recognize and enjoy this little song.

The variations present a different story. It is difficult to write about musical effects and communicate the sound vividly. The reader needs to hear the tune. But a few words on the early variations will suffice. The variations are easy to find and hear, even in places like Wikipedia under "variations on 'Ah vous dirai-je maman.'" The first variation stays fairly close to the melodic theme in the upper voice (right hand), and the harmonic counterpoint is quite similar, as one sees in the performance of the theme. However, a child would have a hard time recognizing the melody and would probably fail to enjoy it. The reason is that the melody is played in sixteenth notes, that is, in notes that are one-sixteenth of the duration of a whole note. It is also played in legato, that is, without any breaks between the notes. This allows the pianist to show off his skills but produces a paradox: the sixteenth notes are very quick, triggering a breathless, intense effect, while the basic melodic notes continue in the original tempo. This is like running 50 meters in 10 seconds but taking 150 steps to do it instead of the more natural 30. The resulting sound is extremely condensed and difficult to enjoy. The clear pattern of the child's familiar tune—the question-and-answer, the arousal and dissipation of tension—are far harder to detect. Our three-year-old girl is not likely to enjoy Mozart's first variation, unless she is musically gifted.

The issue that occupies us here is not what types of aesthetic decisions Mozart wrestled with in order to compose the variations, or what kind of standards may be used to evaluate the artistic quality of each variation. The issue, again, is twofold: what makes the increasingly complex variations pleasing, and what psychological skills must a child acquire in order to enjoy the more complex music? This is precisely the inquiry we make with literature in comparing "The Three Little Pigs" to, say, Dostoevsky's *Crime and Punishment* or Ian McEwan's *Atonement*. We shall not ask what types of literary (or musical) techniques are utilized to create a sophisticated work of art, however interesting it may be. Instead, the question is this: how can a young reader learn to enjoy literature in which, for example, everything takes so long to unfold, where the punishment that mitigates the crime occurs more than halfway through the book, perhaps even beyond?

Delay and Counterpoint

THE THREE EXAMPLES OF food, literature, and music illustrate how the child's sensibility may be trained to go from enjoying simple pleasure to engaging in more complex forms of enjoyment. At the top of such appreciation—or close to the top—we read a story in which the hero is decreed by his father to be sacrificed and dies a horrible death, all for the sake of complete strangers. It's been a bestseller: over one billion Christians love this story.[23] But this is hardly enough—after all, the martyr does not train in order to refine his *aesthetic* sensibilities. Rather, he wants supreme happiness, and I have promised the reader that this chapter will link pleasure to happiness. What is still missing is a description of how training in pleasure helps the child acquire special emotional skills. These skills are the foundation of advanced hedonics (appreciation of pleasure).

Music, our last example, is the perfect place to observe how emotions are trained. Our response to music is, after all, what Plato meant when he said, "This is why we have what we call songs, which are really charms for the soul."[24] But the study of musical complexity and its emotional by-products is a huge field. It may be best to focus on just two basic themes that music training shares with literature and food training: delay and counterpoint.

It is safe to say that music acts as a sound structure unfolding in time with an amazing capacity for emotional evocation. Every conceivable human feeling, from terror to delight, from sadness to awe, can be invoked by music. Taking into account the essential temporality of music, at the most basic register of emotions, from the perspective of evolutionary psychology, are anticipation (or expectation), surprise, and fulfillment. It was the mid-twentieth-century researcher Leonard Meyer who first drew attention, in the 1950s, to the central role of expectation in musical response.[25] More recently, that work was elaborated and updated, with an eye to the evolutionary roots of music and musical response, by David Huron, who is a cognitive musicologist at Ohio State University.[26] Expectation, as we saw in earlier discussions, is the product of successful biological adaptation to novelty. A child who delays the instant response to novelty by judging that it is safe to be surprised or that spicy food can be enjoyed will come to anticipate the new dish. Expectation grows to be a part of the way she encounters the world, and she will ask Daddy to tell the story again—maybe even make it a bit scarier. All of this just builds on simple biological tools we use because proper anticipation of likely outcomes, which is really an arousal that prepares us to act, is essential for survival.

One of the most basic methods of making a tune or a story (or for that matter a dish) complex is to delay the outcome (the simple pleasure of replenishment). If a musical score creates suspense that leads to some resolution but several musical phrases are inserted in between, the listener must learn to retain the effect of the initial phrase but delay the gratification of the resolution while still enjoying the overall melody.[27] If the criminal is not apprehended until after 365 pages of pursuit, and the reader does not peek at the last page, delay becomes a central part of the reading experience. It is usually toward the end of the book that one cannot put it down.

Delay seems to mean just slowing things down. But in the context of adaptation it also refers to effective skills for dealing with new situations that cause stress. For example, a hunter may stumble into the territory of another group and be confronted by a band of fighters. He can respond instantly, probably by turning and running, or he can pause to assess the situation and perhaps try to appease the others. The ability to delay the response involves new psychological tools—for example, mastery over panic, analysis of signals, imagination, and tolerance of ambiguity and stress. A successful outcome produces additional psychological consequences, like a sense of mastery over one's environment and over one's fear; this feeling emerges after the event itself and the initial response to it have passed. Because the delayed reaction can be essential for survival, its results come to be experienced as pleasing (or aversive in the case of failure). The hunter feels deep satisfaction if his mastery over fear allows him to return home safely, perhaps even with his kill. Subsequently, as a result of this learned experience, the hunter can feel the symptoms that accompany fear—such as tingling skin, hair standing up on his neck, slight shivers, gasping—as pleasant. He knows when it is safe to feel these sensations and may actually seek to induce them in other ways—for example, through art, music, games, or rituals.

Musical and literary temporal manipulations can exploit these capacities in order to increase the enjoyment of art. The point is not to educate but to entertain. But the increasing sophistication of the audience requires inventiveness and innovation. As the writer Jane Smiley put it in a *Washington Post* interview on February 20, 2007, even with a book that is saturated with sex, such as her own *Ten Days in the Hills*, the biggest challenge for the author is the reader's boredom. Smiley's own secret is to create narrative tension and a mystifying puzzle that prevents the plot from settling into a predictable course.[28] Music can put similar devices into play. For example, as Huron notes, the psychological impact of temporal delay increases the tension that precedes the musical resolution, thus deepening the enjoyment. Musicologists and psychologists of music have even calculated these responses according to statistical formulas. Beethoven was a master of temporal manipulations (retardation, acceleration, syncopation), and his phrasing can literally make the hairs of your neck stand up—a sign of genius, according to A. E. Houseman's famous quip, and a biological response to a threatening new situation.[29] Glenn Gould reinterpreted the piano music of Bach, such as the Goldberg Variations, which everyone in the early 1950s thought they knew, by animating it with stunning temporal inventiveness. I still get shivers from hearing his performance of the variations, which I have heard countless times, especially the later recording (which is slower than the first). It is unsurprising that, as Huron notes, "the effects of delay in music will be greatest when applied to the most predictable, stereotypic, or clichéd of events or passages."[30] Or, to mix high art with low, everyone knows that the teenage horror movie is about delay in execution (in both senses). The more such movies you have seen, the more you appreciate the director's inventiveness.

These observations about music and film correspond to the idea that the emotional effects of delay are most pronounced when the expectation regarding a particular narrative outcome is most certain. Think of a game of peekaboo with a one-year-old child in which you hide your face behind the corner of the wall. How many times can you show yourself at precisely two seconds? A delay will increase the child's pleasure (if not too prolonged) after two or three times. Try jumping from two seconds to five; the shrieks of joy are born of surprise, a recognition (implicit!) that one has failed to properly anticipate an outcome. The physiological manifestations of this thrill—increased heart rate, laughing, gasping for air, holding the breath, flailing the arms—are the organism's primordial responses of fight, flight, or freeze, characteristic of adaptive situations. The delay in time, when the actual outcome is predictable, makes for a sense of thrill and relief.

You can play around with the pacing of "The Three Little Pigs" as long as your child knows the outcome—but don't try to change the ending! You can add a few changes to the beat of a tune, but your child must be ready for the level of novelty you introduce.

If mere delay and the anticipation it generates do not seem like such a big deal compared with musical inversions, counterpoint, and variations or innovative plot complexities, their emotional effects can be profound. The following anecdote from *Reader's Digest*, quoted in Huron, illustrates this:

> An amateur golfer challenged his club pro to a match. "But," said the amateur, "you've got to give me a handicap of two 'gotchas.'" Although the pro had no idea what a "gotcha" was, he was confident and agreed to the terms. Just as the pro was about

> to tee off, the amateur crept up, grabbed him around the waist and shouted, "Gotcha!" They finished the game without incident, but the pro played terribly and was beaten. Asked why he had lost, he mumbled, "Have you ever played 18 holes of golf waiting for a second 'gotcha'?"[31]

According to some psychologists, delay and the enjoyments it produces may rank among the most consequential aspects of a child's emotional and cognitive growth.[32] The stretched-out expectations and deepened anticipation, both in stories and in music, are not designed to develop greater artistic sophistication by teasing the child. Instead, the delay itself is meant to thrill and satisfy. But it works only as long as it is consistent with the child's mental and emotional abilities. In fact, some theorists argue that delay—the separation in time between conflict and resolution—is critical to the child's proper development, indeed to culture as a whole.

Stanley Greenspan, who is a psychiatrist at George Washington University, and Stuart Shanker, a philosopher from York University, have recently published a detailed theory that argues that the origins of symbolic thinking lie in the exchange of simple information during the interaction between parents and infants. Children learn to interpret the world and respond to it, and do so without fear or anger, through playful circles of communication with their parents. Such feedback cycles "facilitate joy, pleasure, assertiveness, and curiosity, and encourage looking, listening, smelling, touching and movement."[33] From an evolutionary-psychological perspective, managing the child's response to novelty reduces stress, increases pleasure, and promotes further exploration. And all this happens by playing games that turn the child's response to a stimulus into something thrilling. Clearly, then, pleasurable (playful) emotions nurture cognitive development. According to the University of Utah psychologists Dan Messinger, Alan Fogel, and Laurie Dickson, the pleasurable emotions include anticipation, suspense, release of tension, engagement, the excitement of specific activities, the pleasure of watching the caregiver and of giving her pleasure, and the enjoyment of sensory experiences.[34]

Suicide terrorists do not train to write scripts or to gain an appreciation of Bach. But they do undergo hedonic training. What I mean by that is that they train to short-circuit the cycle of stimulus and response and insert elements by which they gain mastery. But unlike the artistic connoisseur, who learns to anticipate and enjoy literary paradoxes, musical counterpoints, or contrasting flavors, the terrorist experiments with fear, anger, empathy, and similar emotions that he seeks to understand and master.[35] If he is successful, he will not only kill a lot of people but also become happy. Special religious orders—in fact, religion in general—also produce states of happiness by interrupting the adaptive cycle through various means of delay and by inserting specific elements between the stimulus and the response. Here too the mastery potential of the adaptive cycle is the key to success, as we see in the following two examples: the writings of the Dalai Lama and the training of Jesuits.

Religion and Happiness

THE PROMISE OF RELIGION to deliver lasting happiness is rarely couched in the language of pleasure.[36] On the contrary, pleasure is often regarded as the obstacle. In a recent bestselling book, *The Art of Happiness*, the Tibetan Buddhist leader, the Dalai Lama, states in unequivocal terms that happiness is the goal of life: "I believe that the very purpose of our life is to seek happiness."[37] It is not something that we do as a matter of our nature; it is what we should be doing. But the Dalai Lama also insists that happiness is not the same as pleasure, which we do in fact pursue relentlessly. Pleasures depend on external circumstances and are fleeting, demanding constant replenishment. Howard C. Cutler, who cowrote *The Art of Happiness*, emphasizes the distinction by discussing the two most notorious features of pleasure: the hedonic treadmill and the susceptibility of pleasure to context. Cutler tells the story of a nurse who came into a huge sum of money only to discover after the brief initial excitement that her level of happiness had not really changed much. In contrast, a friend who was diagnosed with HIV actually reported feeling happier. These are common examples of adaptation to replenishment pleasure—simple response to novelty.[38] And of course, in order to keep up the level of satisfaction from such external circumstances as money or sex, one would need to continuously ratchet up the stimulation—hence the term "hedonic treadmill."

The second example, our dependence on comparison with others for pleasure, is also very familiar.[39] Recall the driver in the traffic jam: if research is correct, people feel somewhat compensated for their slow progress as long as they inch ahead faster than the cars in other lanes.

The key to happiness, according to the Dalai Lama, lies not in improving our circumstances but in changing our mental habits in relation to the external world.[40] This requires a training regimen that is easy to understand, if harder to implement. It consists of the following critical steps:

1. Set a clear and overarching goal for the mental training program.

2. Recognize that happiness depends on positive feelings (compassion, loving-kindness, generosity) and on eliminating the negative (anger, resentment).

3. Analyze with great care how external conditions cause either positive or negative feelings to arise.

4. Develop a firm resolve to control your emotional response to external circumstances. This is the hard part of the training procedures, requiring patience, determination, and courage.

5. Include a variety of exercises such as positive affirmations ("I will utilize this day in a more positive way") to build awareness and strengthen your resolve to develop positive feelings.[41]

In terms of the hedonic-adaptive cycle, the Dalai Lama's teachings are right on the money. The response is separated from the novelty by means of analysis, which contains both a clearly stated purpose (happiness) and a technique for controlling the immediate reaction in order to reach that goal. It is worth noting that the Dalai Lama is not a New Age therapist. His ideas do have religious foundations—beliefs about human nature, a particular understanding of how causes and effects work—but he often engages Western audiences in ways that go beyond either Christianity or Buddhism. One may properly wonder if *The Art of Happiness*, coauthored by a physician, is translating traditional religious goals on behalf of a secular book-buying audience that is bent on becoming happier. Would a religious leader address his followers differently in a monastery, for example? The answer is that there is no reason to make such a sharp distinction: traditional religious training and happiness go hand in hand, and not only in Buddhism but also everywhere else. Even the training of Jesuits supports such a conclusion.

The Spiritual Exercises of Saint Ignatius

A FOUR-WEEK PROGRAM FOR educating the soul, leading it from spiritual rawness to maturity, owes its existence to Saint Ignatius, who learned from his own experiences in Manresa, Spain, in 1522–23. The monthlong training, a spiritual retreat, is far more theological than anything the Dalai Lama has prescribed. Its stated purpose is closely related to the life and death of Christ, its language bound up with a battle against Satan and against sin. No analysis of Ignatius's Spiritual Exercises can ignore the explicit religious intentions of the author, but the practices themselves exhibit a remarkably practical approach to attaining happiness.

In fact, happiness is inseparable from the realization of the theological purposes of the retreat. Close to the end of the retreat, "I will see in mind's eye the contemplation that I am about to make and I will strive to feel joy and gladness at the great joy and gladness of Christ our Lord."[42] The final achievement of the entire month, a life of service to God and realization of His Love, is described as profoundly enjoyable. The four weeks that precede this happy outcome represent a sophisticated practical program. The theological language involved can hardly mask this fact.

The program, which takes place under close supervision, begins with early exercises that appear penitential and even purgatorial, including physical and mental self-chastisement. These can include vigils and prayers, fasting, and living in uncomfortable quarters. Penance implies sin, which has the appearance of metaphysical forces on a satanic battlefield. The concrete reality is more pragmatic. The practitioners must learn to distance themselves from their habitual and knee-jerk responses to external triggers. Rage, hostility, defensiveness, pride, attachment—all of these are "sins," but they also happen to be the "sins" of urban commuters, not just

Jesuit novices. These are also the same destructive emotions about which the Dalai Lama speaks in *The Art of Happiness*. Ignatius is explicit about this: the purpose of these exercises is to help the novice to "conquer himself, and to regulate his life so that he will not be influenced in his decisions by any inordinate attachment."[43] In fact, the text is so psychologically astute that the supervisor is cautioned against overly eager practitioners, even those who wish to embrace poverty or who show great spiritual zeal. There is as much attachment (pride, for instance) in this as in its opposite.

The separation from habitual responses requires resolve, but this depends on honest self-awareness and a subtle understanding of the nature of sin and its causes and effects. The purpose is knowledge, not self-debasement. Separation and understanding serve a clearly articulated goal. The overall purpose of the retreat is made abundantly clear at the beginning of each week: a contemplation of God, service to God, and a desire for His grace. The goal and the path (eliminating attachment) go hand in hand in a profoundly intimate way: "They wish to free themselves of the attachment, but in such a way that their inclination will be neither to retain the thing acquired nor not to retain it, desiring to act only as God our Lord shall inspire them."[44] In other words, the elimination of attachment does not mean making opposite choices (hitting yourself on the head, so to speak), but acting strictly according to the service of God. Translated into existential language, this means that the actor no longer responds to the stimulating event one way or another but gears his actions toward an overarching purpose. The behavioral upshot is moderation and temperance. The emotional consequence is a deep and abiding happiness.

To summarize the general procedure behind the retreat: It begins with a separation between the (sin-causing) event and the (sinful) response, followed by an analysis of the nature of such habitual responses. At the same time a clearly articulated goal directs human behavior away from habitual patterns and toward a single goal. Motivation develops, along with discipline and perseverance. Success results in happiness. The final joy could be interpreted as divine grace because the practitioners have trained themselves to act without attachment—that is, with diminished awareness of their own individual will. But the feeling of divine reward is inessential: that same joy saturates the reports of Buddhist practitioners, who do not believe in God.

Conclusion

We have seen that pleasure is a natural adaptive product and that happiness emerges as a complex form of mastery pleasure. Some readers may insist on one important problem: individuals feel pleasure and evolutionary forces become manifest within individual bodies. Religion, on the other hand, is a group activity. It exists as a church, a chosen people, a sect or caste. But the values of the group, including its understanding of what will bring happiness, often conflict with the pleasures of the individual. If happiness emerges out of pleasure, how is the individual-society gap bridged? How does the group make its members not only conform to the group and even sacrifice their life for the group, but actually enjoy this? This is the subject of the next chapter.

CHAPTER 5

Disgust and Desire: Why We Sacrifice for the Group

You can teach your child to be happy by helping him acquire the habits of self-discipline, setting important goals and thinking positively. Simple pleasures will become lasting happiness for him, even if he is neither Buddhist nor Jesuit. But can the same developmental program apply to a whole society? Does it ever make sense to discuss pleasure and happiness in the context of large groups, or does the picture become so complex that we must abandon simple affects (pleasure or pain) in favor of such things as moral values, law and order, and religious codes? In this chapter, I show not only that pleasure and evolutionary theory can explain the values and actions of groups, but that it accounts for some of the most dangerous—and self-destructive—aspects of religion as well.

The Templar Knight

Novices seeking entry into the monastic-military order of the Templars heard during their admission ceremony that only three things were worthy motivations for joining: escaping the sins of the world, worshipping Our Lord, and doing penance for the salvation of the soul. The role of the Templars, founded in the late eleventh century after the first crusade, was to defend pilgrims traveling to the Holy Land. But as Saint Bernard of Clairvaux put it on their behalf, killing for Christ was not homicide but malecide—the slaying of evil. "To kill a pagan is to win glory for it gives glory to Christ." Dying in such endeavors was martyrdom.[1]

The Assassin

On October 16, 1092, a Muslim called Bu Tahir Arrani, disguised as a Sufi and bearing only a knife, assassinated one of the most important officials of eleventh-century Islam. The victim was Nizam al-Mulk, the vizier (chief administrator) of the Seljuk Empire in the Middle East and Persia. This was the first major strike by a member of the *hashishiyyin* sect, a word that means "those who use hashish" but has evolved into the word "assassin." The Assassins was an elite terror group within a fragment of Shi'ite Islam. Its members were intensely religious volunteers who undertook suicide missions by assassinating high-profile and well-protected officials while using only a knife.[2]

The Hizbollah Volunteer

On April 25, 1995, twenty-six-year-old Salah Ghandour drove a car that was rigged with explosives into an Israeli convoy in southern Lebanon and blew himself up. Ghandour launched this attack against the wishes of his Hizbollah commanders, including Hassan Nasrallah, who thought that the suicide mission was unnecessary. It took Ghandour three years to persuade his superiors before he finally got the green light. His final videotaped statement shows a man going to his death with great joy.[3]

What do these three cases have in common? Obviously, all demonstrate a willingness to kill and to die for a greater cause. All three illustrate the work of people who are assertive, even aggressive, in taking their ultimate fate into their own hands, which represents the outstanding resolve of a religious elite. The three cases also showcase the stunning power that religion and society can exert in motivating people to kill and to die. Now, while the killing is all too common, what about the dying? Why does it look so easy?

Palestinians have been fighting for their cause for nearly a century and using terror at least since the 1950s. Why did they suddenly switch, during the 1990s, to suicide forms of terror? Was it because they had become more desperate, their squalor more extreme? The answer is a resounding no. Imported via Hizbollah from Iran, where it was used extensively in the war against Iraq (1980–88), suicide terrorism took root in the Palestinian territories and Israel only when conditions became right to produce a violent religious elite who would carry it out.[4]

By using the example of the Templars and the Assassins, I am linking the suicide terror in the Middle East today to a broader and deeper context: throughout history and around the world, individuals have left their families and their livelihoods and, disregarding the pleasures to be had in this world, either joined religious or military orders (such as the Templars) or gone into seclusion in remote deserts and forests for the rest of their life. Many of them damaged their health from fasting and the practice of severe austerities, even dying prematurely from the extreme discomforts they brought on themselves. The Templars and Assassins, of course, fully expected to die. Most experts on religion today agree that the ascetics and the members of special orders came from the middle and upper classes of society rather than the peasant, serf, or labor classes for whom life was least pleasurable. Self-denial was not an escape from despair but from prosperity.[5]

Why is this? Why would religion drive the relatively affluent into hard asceticism and self-sacrifice when their lives are fairly easy, even enjoyable? The previous chapter may provide one answer: a central function of religion is to transform mere pleasure into happiness. Those whose lives are hardest are least likely to regard pleasure as an obstacle to happiness, and those whose lives are rich with pleasure (or comfort) but lacking in religious discipline come to feel that pleasure in itself is empty. One has to be lucky in life in order to yearn for something beyond the basics.

But simply renouncing pleasure is not enough. One needs to renounce pleasure for some goal, which is defined by a specific religion and enforced by a society. In fact, that society itself must be defined in religious terms: it has to be a "moral society," one that has no place in it for the likes of one's enemy. Under the thrall of religious hedonics—the promise of true happiness and its attendant membership in the moral society—an individual will commit acts of extreme aggression against his own body and against the bodies of his enemies. In his mind, these acts constitute self-sacrifice and holy war.

All of this behavior has biological roots. However, like all biological causes of complex social behavior (and what is more complex than the oath of the Templars?), these are deeply hidden. How would an evolutionary biologist explain the joyful willingness to regard one's body as a living bomb and to end one's existence for the sake of strangers and in the name of abstract ideas? Getting at the biological—not cultural—roots of something as bizarre and irrational as suicidal terror is as tricky as reconstructing a whole dinosaur from a single remaining tooth. A few biologists have tried to do this sort of thing, most recently Richard Dawkins, and have failed.[6] The hard part is bridging the gap between elaborate cultural activities and simple organic causes. As one critic of biological "reductionism" once said, and as the Marquis de Lafayette famously demonstrated to George Washington during the American Revolutionary War, people go to war not just out of aggression but for honor and even for love.[7] So, to simplify a huge task, I will isolate a single basic—and surprising—theme. I will explain, in leaps and bounds, how food pleasure and eating habits reflect cultural evolution, leading from the earliest hunter-gatherer communities to the suicide terrorists of Hamas and Islamic Jihad.

Why food? We could look at other things, like war, sex, tools, or money. But food is the most universal and basic pleasure that humans enjoy. It is also the foundation of human social existence; it is the essential commodity shared within groups, and it generates the most potent symbols for separating one group from another. Food pleasure is the symbol, the metonym, for human social evolution, and it helps explain how religious societies inspire their members to kill and to self-destruct for abstract reasons.

Food and Evolution

Our ancestors spent most of their waking hours scavenging for food. Opportunistic omnivores, they tried just about everything, from roots and berries to small and large game, including even animals they found dead. Unlike herbivores or meat eaters that were genetically programmed to eat a specialized diet, humans had to make a choice: What should they try once and discard? What should they keep eating? Initially lacking a theory of health and nutrition—even a mythical one—they were guided by one simple mechanism: the search for pleasure. Anything that tasted good was worth swallowing. Food that tasted bitter or smelled foul was avoided.[8] Sweetness was an effective signal for lots of calories, and a fatty taste meant useful fat and proteins. A scavenger can survive and reproduce following such simple guidelines. Early humans also enjoyed the act of eating, and eating until they were engorged. The reason for that was also simple: they could never count on the next meal.

Food was deeply implicated in the evolution of human culture, which is the source of acquired enjoyments and aversions. From the very start, finding food and even eating it required cooperation among individuals. Men and women, both hunting and scavenging while raising offspring, needed to work together. Food, in short, motivated cooperation.

But why should primitive individuals cooperate with others? If we assume that humans were subject to evolutionary pressures, just like other animals, then the driving force had to be the interest of the selfish gene operating in a world of other selfish genes in a way that would maximize its own chances of replicating itself. So why not cheat whenever possible? What would compel members of primitive groups to distribute resources equitably rather than sneaking more meat and other foodstuff to their immediate kin?

Over the last four decades, two views have predominated on the question of how individuals come together into cooperative groups. The first was made famous by Robert Axelrod, who followed in the footsteps of Richard Dawkins in rejecting the notion that evolution operates on entire groups.[9] Call Axelrod's theory the dog-eat-dog theory, or the game theory understanding of cooperation. According to this view, genetic groups (kin-based) display cooperative behavior such as food sharing because, as long as the group is close enough genetically, this fits into the gene's reproductive strategy. Cooperative behavior in itself would not lead to larger groups, such as extended families or primitive hunter-gatherer societies that share food. Another principle had to account for those, and Axelrod dubbed it tit-for-tat, which means cooperation or returning the favor.[10] This is an extremely simple strategy (in the language of game theory) for producing more benefits to the individual than the costs he incurs.

But how would such a strategy get started and gain traction in a world where you could not trust others to reciprocate? What if you gave someone an extra portion of food and next time you did not receive the same consideration? Why would you initiate cooperative behavior if you could not trust others to follow suit? Axelrod and William D. Hamilton answered this with the notion of a simple group called a "cluster."[11] Defections (noncooperative behavior) are characteristic of random encounters in a population of strangers. The more likely you are to meet another individual again and again within a larger universe of noncooperators, the more beneficial it is to begin a tit-for-tat relationship with him. The condition for cooperative groups, then, is the formation of clusters—that is, individuals who meet repeatedly. Once a cluster emerges, there is more than one strategy: either selfishness—so-called defection—or tit-for-tat. This makes the cooperative strategy even more beneficial than before, and it begins to take hold. These clusters represent the beginnings of culture, inasmuch as individuals have to recognize the members of a growing cluster group. Memorizing appearance is one option, but markings and developing group characteristics separated by boundaries enhance this way of distinguishing members of this group from nonmembers.[12]

Some evolutionary biologists have recently denied that this view sufficiently explains the emergence of elementary cooperation, altruism, and group solidarity. David Sloan Wilson and Elliott Sober, for example, have insisted that evolution does in fact operate on the group level by producing behavior codes that bind members of the group into a proper social unit to the advantage (fitness) of the group itself.[13] According to this view, our ancestors distributed food in a strikingly equitable manner and did not cheat because an effective moral system evolved as an aspect of human mental development, in tandem with the group's survival interest.

The two views are as different as Republicans and Democrats, and it is still impossible to say which is correct. According to one, the group initially resulted from a simple selfish-gene strategy that led to clustering; according to the other, the group evolved as a feature of the human mind. The first is hedonic, the second moral. But however they began, groups became critical for human survival. And not only was food the main substance obtained and distributed within the group, but it also became a major tool for marking one group from another. The larger the group, the fuzzier membership became and the more the boundaries had to be fortified in a variety of ways, including through the symbolic use of food.[14]

Pollution and Disgust: Marking Religious Groups

THIS IS A FINE evolutionary tale, but what does it have to do with the complicated world in which we live? What should we make of fifty thousand years of cultural developments, with the rise of farming, husbandry, manufacturing technologies, migrations, and urbanization? And how do these complicated but "hard" features of human societies relate to far more abstract realities such as ideology, tradition, religion, and other bedrocks of modernity? Have these not made the picture of human culture so complex that it becomes impossible to isolate the work of a single factor such as food and one dimension such as pleasure? Some researchers solve such problems by studying simple cultures in places like New Guinea, where food may still act as a currency of social cooperation and as a group marker.[15]

But here is a radical hypothesis: it is possible that by arbitrarily focusing on just the pleasure and displeasure (hedonics) of eating and by limiting this analysis to a reasonably concrete time span, some new bit of knowledge may emerge about infinitely complex social events. There may be only a single thread of correlation running between food pleasure and social complexity. I am fully aware that professional sociologists—and certainly my colleagues in the sociology department at Georgetown University—may think this a bit quixotic. But think of food pleasure as the CHECK ENGINE light on the dashboard and social complexity as the engine itself. The light is connected to the engine by wiring and monitors, but it functions as a simple sign. This may be the extent of the relationship, but who knows?

Before the period on which I focus, food still decisively marked the boundaries of groups. Nowhere was this more telling than in Judaism and Hinduism. In both great traditions, food was rife with meanings, manifested most effectively as rules of pollution and purity. In Judaism food purity still acts as a boundary between the community and its external environment, while in Hinduism food purity marks off one caste from another. Despite millennia of philosophical and theological elaborations within both traditions, the basic fact remains simple: food is a boundary marker.

In Judaism the Torah law books, specifically Leviticus and Deuteronomy, insist on the purity of food:

> You shall not eat any abominable things. These are the animals you may eat: the ox, the sheep, the goat, the hart, the gazelle, the

> roe-buck, the wild goat, the ibex, the antelope and the mountain-sheep. Every animal that parts the hoof and has a hoof cloven in two, and chews the cud, among the animals you may eat. . . . And the swine, because it parts the hoof but does not chew the cud, is unclean for you. Their flesh you shall not eat, and their carcasses you shall not touch. (Deuteronomy 14:1–8)

Everyone knows that Jews who keep kosher do not eat pork or shellfish and a host of other foods and that they famously avoid eating meat and milk products at the same time. Most of the remaining details are unfamiliar to most people. Still, theories abound for rationalizing such rules, particularly hygiene (the vulnerability of pork meat to trichinosis). Throughout the centuries, both rabbinic and secular scholars have interpreted the exquisitely detailed dietary rules in a variety of ways, ranging from moral symbolism to ancient sacrificial taboos, contemporary physiological theories, and so forth. The most persuasive sociological theory remains the theory of Mary Douglas, which was published in 1966 in her seminal work, *Purity and Danger.*[16] According to this idea, holiness (the purpose of the laws) means separateness. The laws of food are about marking and keeping boundaries. The most obvious boundary is between the Israelites and the surrounding nations that have been "cut off" by God and are on the verge of being dispossessed (Deuteronomy 12:29). The Law-giver cautions the Israelites to avoid imitating the idol worshippers in any way, but especially in their worship and eating customs. But the dietary rules also display a type of classification, or a traditional way of defining species: chewing cud, split hoof, having fins and scales, and so forth. Creatures that seem to transgress the dividing line between species—for example, those that live in the ocean but do not have fins and scales (shellfish)—are regarded as anomalous and are therefore abominations: they are impure. They are like the two-headed calf at the county fair—monstrous, repugnant, and inedible.

There are other principles, of course. The avoidance of mixing meat and milk products introduces an ethical dimension: meat must be properly slaughtered, and anything resembling the food habits of the pagan neighbors must be rejected, if for that reason alone. As food rules expanded and became more systematic centuries later in the Mishnah and Talmud, which devoted entire tractates to the topic, the principles continued to evolve. But the notion that food purity was related to holiness remained central. And being holy meant belonging to the group and owing allegiance to the best ideas that guided the group. For example, the rabbinic school of Shamai insisted that a man could sell his olives only to his associate—a strict reading of what it means to belong to the group. The competing school of Hillel, which was more lenient, argued that a man could sell his olives to anyone who merely paid tithes.[17] Either way, the food transaction remained internal to the community.

Hinduism is just as famous—some would say notorious—for its laws governing the purity and pollution of food. It is a familiar fact to anyone who has studied Hinduism that Brahmins will not eat in a restaurant where the cook is anything but a high-caste Brahmin. Like the Jew who insists on a strictly kosher kitchen, anything less than pure food will not do. The Hindu rules governing what is pure food and which foods are acceptable in the kitchen are almost as detailed as the biblical rules. One classical source of food rules is the ancient law book of Manu (Manu Smriti).[18]

The rules in Manu do not come with an introductory preface, like the biblical one about holiness and separation. The authors simply jump in and list dozens of foods that a Brahmin (the purest caste and the target audience) must avoid, including garlic, scallions, onions, mushrooms, the first milk of a newly calved cow, spice cake made of flour, butter, and sugar, the milk of a camel or ewe, carnivorous birds, whole-hoofed animals, village birds, goose, birds that dive and eat fish, and many others (Manu 5.1–26). Brahmins are known for avoiding meat, and Manu is very detailed about the rules that do allow meat and the ethical principles that prohibit it. For example, the prescription "You may eat meat that has been consecrated by the sprinkling of water" (5.27) would cover the eating of sacrificial food, as long as it is properly purified by ritual. On the other hand, a broad principle stipulates that "he whose meat in this world do I eat will in the other world eat me" (5.55).[19]

Ethics aside, it is extremely difficult to detect the logic of food classification in such rules. Does Mary Douglas's boundary marking still apply here? A clue is provided by another set of more general rules pertaining to food, in chapter 4 of Manu:

> He should never eat the food of those who are drunk, angry, or ill, nor food in which hair or bugs have fallen, or which has been intentionally touched by the foot; nor food which has been looked at by an abortionist, or touched by a menstruating woman, or pecked at by a bird, or touched by a dog. (4.207–8)

In other words, a pure Brahmin may not eat food that has been contaminated by impure objects, animals, or people. Eating makes you vulnerable to pollution, the very substance that the caste system is designed to minimize by means of separation.[20]

Scholars have not agreed on what exactly constitutes pollution, either in Judaism or in Hinduism, or for that matter in other religions as well. Corpses are polluting, as is cooked food left uneaten, shoes and unwashed feet in the mosque, menstruation and all bodily waste. Even modern readers, guided by their nose and gut feeling, will get this. But pollution is also something intangible, like having your uncle pass away in another country or eating a succulent pork chop or T-bone steak (if you are a Jew, Muslim, or Hindu, respectively). Sociologically speaking, pollution is the outcome of violating the rules that regulate boundaries. Anything that crosses a boundary is polluting, whether that boundary separates the inside of the body from the outside, or my body from those of others, or one caste from another, or one nation from all others.

These boundaries are surprisingly powerful and generate intense emotions. When my students doubt this, I offer them the following experiment. Take a clean cup, one that you have just washed. For the next ten minutes, instead of swallowing, which you do constantly, spit into the cup. In ten minutes the cup will be half full. Next, drink your own spit down. Go ahead. What is your reaction? My students respond with outbursts of disgust. "Ugh," they grunt, contorting their mouths, wrinkling their noses, sticking their tongues out. Why can we swallow something without a thought but feel repelled by the same substance as soon as it leaves our body? This is a stark example of how powerful boundary-crossing remains, even today. My students feel the same way about exchanging socks with their colleagues (gloves are fine), and even those who let their own dog lick their face would not agree to let someone else's dog do the same.

In short, boundaries, which are symbolic cultural constructions to mark the group, create powerful visceral emotions, and these guard the integrity of everything that belongs inside from outside invasion.[21] And food, along with attendant activities like cooking, eating, digesting, and eliminating, represents the most primitive and still most effective group-boundary guard.[22] If you want to demonstrate and enforce your hatred of another group, you depict them in revolting (boundary-crossing) terms. The Nazis, for example, in their semiofficial propaganda film *Der Ewige Jude*, characterized Jews as rats, creatures that infest garbage and sewers and then cross into the homes of good people to infect them.[23] Martin Luther, the German reformer and anti-Semite, preceded the Nazis. Remarking on the pornographic "Judensau" images that were widespread throughout Renaissance Germany, he wrote:

> Here in Wittenberg, on our church, a sow is carved in stone. Some young piglets and some Jews are suckling her; behind the sow is a rabbi. He raises the sow's right leg, with his left hand he pulls out his member, leans over, and diligently contemplates, behind the member, the Talmud as if he desired to learn something very subtle and special from it.[24]

The anti-Semite here reviles those whom he hates by having them cross a disgusting boundary (bestiality) with the very creature they find disgusting, the pig. Hatred gets a visceral boost from food and sexual taboos. In India, when Hindus wish to outrage Muslims, they parade a pig in front of a mosque. The Qur'an calls Jews pigs and monkeys.

But food-generated disgust is not just about rejecting or hating others. It is a far more complex emotion. In Rohinton Mistry's book *A Fine Balance*, which is critical of the Hindu caste system and its rigid boundaries, purity and pollution become a matter of ridicule. One village Brahmin, the purest of the pure, is described as a flatulent nose-picker and sputum-spitter. In another section, high-caste vegetarian students living in a hostel discover a sliver of meat floating in a pot of lentil soup.

> The news spread, about the bastard caterer who was toying with their religious sentiments, trampling their beliefs, polluting their being . . . within minutes, every vegetarian living in

> the hostel had descended on the canteen. . . . Some of them seemed on the verge of a breakdown . . . poking fingers down their throats to regurgitate the forbidden substance. Several succeeded in vomiting up their dinners.[25]

In Vikram Seth's monumental *A Suitable Boy*, a disgusting old man confronts Kedaranath, who is the son of one of the main characters in the novel:

> The old man moved toward him and held out his stained and dripping hand: ". . . We don't like the Jatavas, we are not like them, they eat the meat of buffaloes. Chhhi!" He spat out a syllable of disgust. "We only eat goats and sheep."[26]

Jewish humor is known for self-deprecation—or more precisely, for taking the barbs of anti-Semites and turning them into jokes. The film *Borat* recently exposed this side of Jewish humor to the whole world, particularly in the scene where the two elderly Jewish owners of the inn where Borat rents a room miraculously transform themselves into cockroaches and invade his room. Grigory Bogrov, the Russian-Jewish author (1825–87), reports on the following exchange between a Jew and non-Jew on the subject of eating meat fried in butter:

> "It's disgusting and nasty. It's treyf. Yuck, how treyf!" I said.
>
> "And kugel, and onions, and garlic—what, they're not disgusting and nasty? They actually stink!" Olga stated with no less aversion.
>
> "No, that's kosher, so it tastes good."[27]

If you are wondering what "treyf" means, a recent bestselling book, *Born to Kvetch*, offers a definition:

> Treyf originally referred to animals that would otherwise have been kosher and was the biblical equivalent of "roadkill." It was later extended to all forms of un-kosher food, and is the model on which such deliberately ironic terms as shikse are based—something attractive but illicit is given a disgusting name in order to make it less alluring. "Thou shall not eat road-kill" is less of a challenge to your willpower than "hands off the porkchops" or "step away from the lobster bisque," and the term treyf is a deliberate attempt to change the essence of that T-bone steak.[28]

Of course, anyone who has read Philip Roth, and especially *Portnoy's Complaint*, knows that just calling someone something disgusting like a shiksa will not keep a Jewish man away—and may even encourage him.

On the darker side of food and disgust, we should bear in mind that the feeling of disgust amplifies and energizes hatred. In that same Goebbels brainchild *Der Ewige Jude*, the image of a Jewish man was superimposed with that of an African—seeking to identify Jews, in the lips, nose, and ears, with the disgusting visual marks of inferior groups. And indeed, the racism of Jim Crow's America, though comprehensive enough to cover every aspect of life, seemed to emphasize disgust. The rules insisted that black people could not touch whites, eat with them (unless separated by a partition), or exchange seats. Blacks could share neither restrooms nor water fountains with whites. The impression is unavoidable that racism is not merely about economic and social subjugation—it is also about the disgust that is generated by the pollution attending the boundary between two species.

It is important to note that Moses and Manu, the mythical legislators of Judaism and Hinduism, did not invent the rules governing food behavior and pollution. We can find similar rules in a wide variety of societies around the world today, but most of these rules are unfamiliar because they are tribal. And herein lies the reason Judaism and Hinduism insisted on food purity while Christianity and Buddhism did not—a reason stemming from a new social and religious arrangement.

Disgust Becomes Restraint

RELIGIOUS TRADITIONS THAT MARK their boundaries with rules about purity and pollution tend to be hedonic. Pleasure is fine. Nothing in the notion of treyf or in the feeling of disgust at other people's food precludes the sheer enjoyment of one's own. On the contrary, the Hebrew Bible celebrates food. The Holy Land flows with milk and honey, and even God encourages dining pleasure: "Only listen to me and you will have good food to eat and you will enjoy the fat of the land" (Isaiah 55:2). Like sex, food in its place is a source of enjoyment, in both Judaism and Hinduism.

But centuries and hundreds of miles apart, food purity lost its power as a group marker when other concepts emerged in the domain of food: desire and sin. As Christianity and Buddhism began to take shape, staking new positions in conscious revolt against Judaism and early Hinduism, new criteria were used to mark off the inside of the nascent group from its outside. The very idea of community changed from the more organic Hindu caste society and Jewish national-tribal group. These were replaced with missionary religions practiced in faith-based voluntary associations with fuzzy boundaries—the Buddhist "sangha" and the Christian "church."[29] One of the first items jettisoned was the notion of disgust as border sentinel, as illustrated in the following Buddhist example.

Liz Wilson writes about a Buddhist saint whose story is told in the Mulasarvastivada Vinaya. A former high-caste Brahmin, Mahakassapa received a bowl of rice-gruel from an old woman.[30] She had obtained it by begging. In the food he found the old woman's leprous finger—it had fallen in when she tried to fish out a fly. Like most of us, any Brahmin would have recoiled at the thought of eating this food, but Mahakassapa ate it and in doing so elevated the spiritual standing of the old woman, sending her eventually to a heaven where, in her heavenly body, she would enjoy heavenly ambrosia. The Buddhist saint turned disgust into virtue, something that most people (except someone like Saint Francis of Assisi) could not dream of doing. The idea is clear: what counts in evaluating food—but really people—is morality rather than rules of purity. This is precisely the idea espoused in the New Testament against Jewish dietary rules. In the Gospel of Matthew (15:11), Jesus declares that it is not what goes into the mouth that defiles a person but what comes out of it. The important thing is to avoid evil intentions, murder, adultery, fornication, theft, false witness, and slander (Matthew 15:17–19).

The point is not that Judaism and Hinduism sanctioned immoral behavior, but that they used other criteria for defining what constituted a Jew or a Hindu. The new religions announced new ways of determining group identity: faith and morality, or correct belief (orthodoxy). This was a major leap in cultural evolution: relatively organic groups became ideological communities. The new social-religious arrangement resulted in a hedonic revolution in relation to food. Food remained important, obviously, but eating suddenly became a moral temptation, an opportunity to sin, rather than a source of pleasure (and threat of pollution). Dining pleasure became literally dangerous. Virtually all the major Church fathers from Paul onward, including Clement, Tertullian, Origen, Eusebius, Jerome, and Augustine, warned against the dangers lurking in food and in dining. Eating, after all, was more common than sex and came before sex in humanity's first sin when Adam and Eve tasted the forbidden fruit.

One notes with amazement that men of entirely different backgrounds—some urban, others not, Greeks, Romans, Egyptians, born Christians, converts, intellectuals, plain speakers—wrote about this topic with a high level of uniformity. The central theme, as Tertullian (c. 160–225) put it, was this: granted that eating is necessary for existence and indulgence is an evil and possibly addicting ("For your own belly is god, and your lungs a temple, and your paunch a sacrificial altar, and your cook a priest"[31]); what makes food a threat is not its necessity but the pleasure that it gives. That is precisely what makes humans beastly as they begin to live in order to eat.

Clement (c. 150–225), who was a contemporary of Tertullian's and a cosmopolitan resident of Alexandria, wrote extensively about food habits. In *Christ the Educator*, he insists that eating should not be our main occupation, nor pleasure our chief ambition.[32] Clement's work reveals the experience of a gastronomic connoisseur and regular diner. He lists the great delights of second-century Alexandria: Sicilian lampreys, eels from Maeander, kids from Melos, mullets from Sciathos, Pelordian mussels, Abydean oysters, sprats from Lipara, Montinean turnips, beets from Ascra, Methymnian scallops, Attic turbots, laurel-thrashes, Persian figs (golden brown), fowl from Phasis, francolins from Egypt, peacocks from Medea. These foods are fried in sizzling frying pans and served up (the fowl is) in sweet sauces, followed by sweets and pastries and honey cakes and desserts. All of this Clement describes exquisitely in order to condemn it: "We must restrain the belly. I know the beguiling lure of dinners."[33] Fight the pleasure of sumptuous meals, he urges, by keeping the food "plain and restrained."

Some Church fathers drew an explicit connection between food pleasure and sexual desire. Saint Jerome (c. 340–420) believed that food and sex were linked by medical science, which he misread in the works of Galen. Food can produce heat, which triggers sexual desire: "Secondly, in the way of food avoid all heating dishes. I do not speak of meat only . . . but with vegetables also anything that creates wind or lies heavy on the stomach should be rejected. . . . Regard as poison anything that has within it the seeds of sensual pleasure."[34]

But by far the greatest expert on pleasure in all its forms, and probably the most introspective early Christian writer, was Saint Augustine (354–430): "It is not the impurity of food that I fear but that of uncontrolled desire."[35] Book 10 of the *Confessions* represents a very precise study of pleasure, with food as a main target. "The necessity of food is sweet to me, and against that sweetness I fight lest I become a captive."[36] Sounding like the addict that he felt himself to be, Augustine adds: "I wage a daily battle in fastings often 'bringing my body into captivity' (I Cor. 9:26–7). My pains are driven away by pleasure. For hunger and thirst are a kind of pain, which burns and can kill like a fever, unless the medicine of sustenance brings help."[37]

All of these writers were deeply familiar with the Jewish insistence on food purity and also with the biblical celebration of dining pleasure. They also had experienced Greek and Roman hedonism, culinary as well as sexual. So why did they suddenly shift? Why did new, negative hedonics emerge during the first few centuries of the Christian era? This is one of the most complex questions in the history of religion. The answer involves new beliefs and philosophies and a complicated synthesis of Greek, Middle Eastern, and Jewish ideas. But there were other, more sociological factors, such as urbanism, individualism, and increasing geographical mobility. A Jewish writer, Philo (20 BCE–50 CE), who did not share the young Church's faith in Christ, underwent a similar hedonic reevaluation owing to his reading of the same Greek literature and living in the great city of Alexandria.[38]

Given such a complicated picture, how can we decide what was more influential—the intellectual factors or the sociological ones? One way is to compare the Christian transition in food hedonics to what we find in India several centuries earlier. There too a radical reevaluation of food took place. The Buddhists rejected the Hindu rules governing food purity, and at the same time they turned their backs on the pleasures of dining in favor of blandness.

The early scriptures of Buddhism, the Vinaya texts associated with Theravada Buddhism, concerned both lay members of the growing Buddhist community (*sangha*) and the core members—the monks and nuns (*bhikkus* and *bhikkunis*). The subject of supreme joy or happiness was ever-present in this discourse—usually conveyed in the words of Buddha—and it shaped the way in which lesser pleasures, such as eating, were conceptualized.

The subject of food and related topics—the quantity and quality of food, the company one kept while eating, the manner of obtaining food, eating times—were central to the lives of Buddhists, both lay and monks, and were entirely instrumental to the goal of obtaining the highest good—supreme and lasting happiness.

According to the Samyutta Nikaya 3.13, in a section called "Donapaka Sutta," food restraint is extremely important. The text tells of a king (Pasendi) who stuffs himself with food before approaching the Buddha. Observing this, Buddha recites the following verse:

When a person is constantly mindful,
And knows when enough food has been taken,
All their afflictions become more slender
They age more gradually, protecting their lives.[39]

Food was also a central feature in how the lay population interacted with monks and in the monastic discipline of those monks who were serious about the supreme goal—*nirvana*. The monks relied on ritualized begging among the villagers as a way of obtaining sustenance and giving back spiritual benefits in exchange. Such an arrangement was carefully regulated by the many texts authored during the first few centuries of Indian Buddhism. Violation of the rules resulted in a specific kind of sin called *pakittiya*. The following extract spells out a few examples of this sin, touching on both the quality and quantity of food:

> Whatsoever Bhikku [monk] when he is not sick, shall request for his own use and shall partake of delicacies—to wit, ghee, butter, oil, honey, molasses, fish, flesh, milk, curds—that is a pakittiya (sin).[40]
>
> Whatsoever Bhikku, when he has once finished his meal, though still invited (to continue eating), shall eat or partake of food that has not been left over, whether hard or soft—that is a pakittiya. . . . Furthermore: A Bhikku who tries to persuade another Bhikku to eat non-left over food after he has finished his meal is also committing a sin.[41]

None of the instructions regarding food in early Buddhism deal explicitly with the subject of pleasure or hedonics as did the early Christian texts. Furthermore, Buddhism is famously moderate, seeking the middle road between pleasure and pain. So while fancy foods and large feasts are discouraged, so are extreme forms of fasting. Still, when King Milinda asks Nagasena why a man should bother becoming a monk if laypeople can also proceed on the Noble Path, the sage answers that, while the layperson should practice moderation because gluttony can lead to unethical behavior and destructive emotions, the monk eats while actually perceiving the repulsive nature of food.[42]

Food and Authority

WHAT DO WE LEARN following such a long discussion of food and the pleasures of food? The most basic point is that attitudes toward food, especially the enjoyment of eating, shift when complex societies become even more complex. Food pleasure reflects social evolution. In the cases I have described, the move was from so-called organic religious groups (Judaism and Hinduism) to missionary or ideological religious groups (Christianity, Buddhism, Islam). The general rule can be stated in the following terms: the more ideological the foundation of a religious group, the greater its emphasis on restraining food pleasure. The rule applies to other forms of pleasure (most specifically sex), but food makes the point.[43]

Why is this? What is the connection between the social organization of a great religious tradition and its approach to dining pleasure? Recall that I began this chapter with a discussion of the forces that hold groups together in a natural world dominated by the selfish gene. Groups depend on the sacrificing of selfish interests for the sake of the group—that is, on a certain degree of cooperation and altruism. The more "natural" the group, the easier it is to sustain the clustering. The looser and larger the group, the harder it becomes to promote social behavior. As a result, the second type of group needs to promote the values of self-restraint in relation to all things hedonic, while the first group needs only to protect its boundaries.[44] Negative hedonics, the restraint on simple pleasures like eating, is not nearly as important within such natural groups. Another way of putting this is that altruism and cooperation in natural groups are easier to achieve. This distinction becomes obvious in the writings of early Christian martyrs (notably Perpetua and Felicitas, who died on March 7, 203 CE), who not only sacrificed their lives for the new Christian community but often showed contempt for and destroyed their own families in the process.[45]

But readers may still wonder how this connects to the topic of the chapter, namely, killing and dying for the sake of religion and the group. How does food-restraint translate into willful suicide? The two types of religious groups I am discussing can be called the "natural group" and the "moral society." The following table illustrates the key differences between the two. Readers should note that these are differences in degree rather than absolute dichotomies, and that they reached their peak of influence by 1300 CE, after which modern developments began to erode the distinction. Furthermore, Buddhism developed in radically different directions compared to Christianity and Islam (no inquisitions and heresy trials).

Natural Group	**Moral Group**
No founder	Founder
Correct practice	Correct belief
Pluralism	Schisms
No inquisitions	Inquisitions
Little heresy	Heresy
Excommunication	Executions/Excommunication
Pollution	Sin
Boundary is key	Center is key

This list of characteristics demonstrates basic sociological facts about the two types of religious groups. In the moral society the center matters above all.[46] This consists of a basic religious doctrine reinforced by a judicial authority. The natural group has no such axiomatic idea, not even the unity of God in Judaism or Brahman in Hinduism. These religions are not orthodoxies but orthopraxies: they emphasize correct practice as defined by exquisitely detailed laws (Torah and Dharma). Furthermore, Hinduism and Judaism rarely if ever execute people—heretics—who go against enshrined religious beliefs. Where they gave birth to distinct forms of practice (such as the devotional groups of medieval Hinduism), these were not the sharp and at times violently divisive practices found in Christianity and Islam and established by people who believed that they knew best what the founder had said or meant.

I should add that all major religious traditions in the world developed within their general boundaries something I call "hyper-moral" communities.[47] These would be monasteries, small devotional groups, and sects or similar voluntary associations that center on an idea and a founder. These hyper-moral groups rarely have the enforcement mechanisms that the tradition as a whole enjoys, but they compensate by reinforcing their own external boundaries: you must take difficult vows (silence, poverty, chastity), pass arduous or painful initiations, or abandon your family and friends.

Moral and hyper-moral communities generate extreme anxiety at the margin.[48] Individuals who find themselves near the border, whose beliefs are questioned (think of non-Catholics during the Spanish Inquisition), have to find their way closer to the center, avoid the consequences of their errors, or leave that part of the world. But here is the devastating psychological fact: because moral communities define the center in terms of moral belief, those who live on the edge come to experience themselves as sinners. Moral communities teach moral introspection by means of practices like confession, and they demand that marginal individuals "return to the table" (Augustine's metaphor) by purifying their hearts of sin. They must confess and perform penance. The result can be dangerous.[49]

When you combine a religious ideology that teaches that pleasures are bad and must be curtailed with a feeling at the margins of the tradition that one has sinned and must perform penance in order to regain admission, self-destruction becomes likely.[50] The person at the margin feels morally unworthy because of his social standing. In practical terms, examples of such a person might be a man who cannot produce children, a woman suspected of adultery, or a professional woman who opts for a career above marriage. They experience their ostracized state as a personal moral failing, as shame or dishonor. And because moral communities promote hedonic restraint as the unifying virtue, the sinner comes to experience his or her body—the locus of pleasure—as the enemy. To find his way back to the center, the sinner must torment his own body: many penitential acts in Christianity and Islam involve actual or symbolic self-flagellation. Pain becomes power, the ticket back in. Those who survived the Inquisition's complex process finally gained admission to good society through their own pain and humiliation.

Conclusion

A MAN MAY SEEK to join the Templars or commit suicide for the Hizbollah without having sinned or having been pushed to the margins of his community. True, a very high percentage of Palestinian suicide bombers do come from among those whose moral standing has been questioned, such as adulterers, single women, suspected collaborators, or sons of men who lost their honor; in these cases, suicide terror is penitential in the sense discussed earlier. It is sufficient, however, that the volunteer merely *feels* morally inadequate, that he wishes to make a name for himself at the very heart of the moral community. The glorious self-sacrifice—the martyrdom—of such individuals thus becomes emblematic of a religious society that calls for sacrifice to hold the center together.

The reason that not all religious Christians and Muslims self-immolate in this way is that there is a clear distinction between hedonic restraint and violent penance. Those who find restraint insufficient, who insist on self-destruction, are often misguided about this important distinction. And what makes matters far worse is that such actors are often motivated by the loftiest religious ideals: faith and spiritual love. That is the subject of the next chapter.

CHAPTER 6

God's Love and the Prozac Effect

IMAGINE THAT YOU WENT TO a doctor and complained about feeling unwell. After running some tests, the doctor ignored the results and told you that you were perfectly healthy because he knew that was what you wanted to hear. Or imagine that your mechanic told you the same about your car—either way, how long would you keep the same physician or mechanic?

Surprisingly, many of us say—or even more strikingly many of us believe—precisely those things that we wish were true. The most obvious example is sports. A few years ago, ESPN.com had a nice feature on its Web site: through polling software, fans could predict which college or professional team would win an important game, a tournament, and so forth. After voting, participants could see the geographical distribution of the votes. So, for example, if you were voting on a Lakers-Celtics game, you could see how all the votes broke down geographically. Predictably, basketball fans closer to Los Angeles overwhelmingly indicated their belief that the Lakers would win, and closer to Boston voters believed the Celtics would prevail. At the very center of the country—say, Iowa—the vote was far more likely to split evenly. Note that fans were asked to predict which team would win, not which team they wished would win—but statistically it seemed to come down to the same thing.[1]

But of course, you say, this is sports—people are by definition emotional when it comes to sports. What about politics? If you turn to the political blogs, such as those published by the *New York Times* or the *Washington Post*, you find that people frequently analyze and predict political outcomes in a similar fashion. A recent PostGlobal posting asked WashingtonPost.com readers to comment, on the sixtieth anniversary of the founding of the state of Israel, how likely it was that Israel would survive to one hundred. Well over one hundred comments were posted, which I read carefully. They split about half and half in their predictions: half the readers predicted that Israel would make it, and half argued that it would not survive. In over 90 percent of these comments, there was a direct correlation between the writer's attitude (sympathetic and supportive versus hostile and critical) toward Israel and his or her prediction: not a single writer who was sympathetic to Israel predicted that it would not survive, nor did a single hostile writer predict that it would survive. The only exceptions were those writers who did not bother to make a prediction and just vented.

This may seem like a nonscientific and impressionistic reading of a pervasive tilt in thinking that runs through our society. But scientists do have a name for it—unconscious confirmation bias—and it has been validated by fMRI brain scans at Emory University.[2] The reader may have noticed this bias at work too. For instance, cable television political shows naturally promote one-sided political analysis, but on these shows what is taken as *prediction* (expected outcome) is rarely separated in a clear manner from *advocacy* (desired outcome). We see this in newspaper op-ed columns as well—writers with a distinct political perspective predicting the very same events that they desire. This bias was most blatantly in effect before the war in Iraq was launched, when advocacy and wishful thinking merged entirely into prediction. No supporters of the war argued that it would go badly, and no opponents of the war (on moral or diplomatic grounds) predicted that it would be a cakewalk.

All of this may be granted even without the sort of scientific measurement that would solidly prove the point. Indeed, I would forgive the skeptical reader who may feel that I too want to believe that such is the case because it bolsters some essential point that I am setting up. Touché—we do become invested in our work. But even a sympathetic reader may argue that sports and politics are notoriously slanted by emotion, and that in no other areas are we so emotionally partisan and so blind to our true motives. On the surface this may appear to be true: it is certainly easy to make the case for human irrationality in the domains of sports and politics, which thrive on division and emotion. But I am going to go beyond merely putting down sports fans and political hacks and make a stronger claim: there is no sphere of human activity or human thought in which thinking based on desired outcomes or on anticipated enjoyed outcomes is more pervasive than in religion. We shall look at this claim in great detail when we explore the concept of heaven later in the book. But the starkest example is the idea of God, an intellectual creation of staggering hedonic proportions. God is the very embodiment of what matters most to us—happiness, meaning, justice, eternal well-being. Even at His worst, God rages mostly against those (Job and a few others excluded) who violate His explicit understanding with us. In this chapter, I show that even the loftiest understanding of God—as the embodiment of supreme love—is predicated on the unconscious joys of a wonderful feeling.

Don't get me wrong: on some level, I sympathize with those who engage in such thinking, and I do so because religion is positively implicated in some of our most painful feelings—as a remedy for the grief of loss.

The Consolation of Faith

On July 3, 2008, my father, who had just turned eighty-seven, was buried in the cemetery of Kibbutz Yif'at, Israel. He had lived a life of great pain owing to an injury sustained in World War II. Despite his mature age, my father's funeral was a very painful event for the family and for many of the friends who gathered on the hill that overlooks the Yizrael Valley. As the oldest son, I read the Kaddish text in Aramaic. The English translation reads as follows (in part):

> Glorified and sanctified be God's great name throughout the world which He has created according to His will.
>
> May He establish His kingdom in your lifetime and during your days, and within the life of the entire House of Israel, speedily and soon; and say, Amen.
>
> . . .
>
> Blessed and praised, glorified and exalted, extolled and honored, adored and lauded be the name of the Holy One, blessed be He, beyond all the blessings and hymns, praises and consolations that are ever spoken in the world; and say, Amen.

This religious text appeals to a creator, a redeemer with his blessings and consolations—but none of that moved me.[3] It had no effect on my sorrow, I suppose because I lack all faith in what the Kaddish claims and promises. The ritual did console me, however, in other ways. My father is buried with my grandparents and great-grandparents, founders of the kibbutz where I had been born, on a hill overlooking the historic Yizrael Valley. The broad, expansive view ends at Mount Carmel just above the Kishon River, where the heroine Deborah fought the Canaanites. In other words, my father has now merged into the stream of ancestral narrative that defines Jewish history in the Land of Israel. In some sense, he can no more be forgotten than the Bible, with its heroes, or the land itself and those who rebuilt the Jewish homeland in this place. I find this consoling in a way, without reference to a supernatural or transcendent power, a God.

Others at the funeral—my sister, for example—did find the prayer consoling. Those who believe in a God that created a good world and promises redemption to the righteous may be less frightened or saddened by the death of a parent. In either case, both the believer and the nonbeliever are motivated to relieve the pain of loss during funerals. But we are not conscious of owing our faith (or lack of faith) to the fear or to the sadness. Instead, we quietly and mysteriously adhere to satisfying ideas: some are historical and national (mine), and others are metaphysical (my sister's). We think of these as meaningful ideas and let their comfort—their hedonic value—do its work unobtrusively.

This subtle hedonism should not be regarded as a denial of God's existence. The fact that the idea of "God the Creator and Savior" is hedonic—a topic that has drawn highly publicized ridicule in recent years—has no bearing on the existence or nonexistence of whatever one means by "God." In the recent debates between believers and the "new atheists," *both* the atheists and the defenders of God appeal to the hedonic consequences of their positions, namely, which outcome is ultimately "better" for us (makes us happier, healthier, more peaceful): the sovereignty of God or of

Man.[4] Frankly, I don't really care. I do not believe that the question of God's *existence* can or should be settled by whether He consoles, redeems, or gives other forms of pleasure. This chapter is about something else: how does the idea of God, or faith in God, make people feel better, and is this a good thing?

When Barack Obama in a campaign speech appeared to attribute people's faith in God to the bitterness of economic misfortune, a flood of angry responses reminded him that, no, their faith was motivated by love and hope—not the desire to escape misery. For the many Americans who claim to be believers, and for contemporary theologians as well, God is not simply a biblical character or a political device, but a living presence that could best be described as transcendent love (or Love). Here is how Paul Tillich describes spiritual love: "This affect is an expression of the participation of the soul in the divine self-love. The courage to be is possible because it is participation in the self-affirmation of being-itself."[5] There is nothing here about smiting enemies or parting the sea. God is not a personal actor or even a subject, but the very ground of being. The atheist's straw man/god fails to account for the primacy of love in the way that both theologians and common people often feel and think about God.

Why does all of this matter? In this chapter, I argue that this love that religious people interpret as coming from God, and on which they often come to construct their view of God, is a form of pleasure. Readers should know by now that I define pleasure in a technical (evolutionary) way. As we know from chapters 2 and 4, which dealt with the subject of pleasure in detail, pleasure is not simply the feeling one gets from something like eating ice cream or driving a new car.[6] Even so, the claim that love (by God for us) is a form of pleasure is audacious and carries strong implications. For example, it implies that this idea of received love—the foundation of sophisticated theologies—can be explained by means of evolutionary psychology, just like all other forms of pleasure. Furthermore, such hypothetical love—it is not truly a feeling but an interpretation of pleasure—can lead to consequences that are not always salutary or consistent with the way we understand love. This is explored in the next chapter on the master and the disciple.

To summarize: the love that people believe is coming from God (and their understanding of God based on this belief) derives from a kind of pleasure that we only interpret as love. And the source of this pleasure is completely natural. The pleasure I am talking about is mostly a special type of exploratory pleasure. Recall that following replenishment or novelty pleasure is mastery, which religion teaches as a method to free us from enslavement to the first. Following on the heels of mastery, in the religious life, is a liberated and even playful joy that belongs in the category of exploratory pleasure. What I intend to do in this chapter is explain how this pleasure emerges, how it is interpreted as love, and how this love is attributed to God, then consider the broad consequences of this process, which is a sort of "hedonic error." To follow this very difficult argument clearly, I must first explain the mechanism of the error, which I have called "the Prozac effect"—following Peter Kramer, who does not actually use the term but describes the phenomenon.[7] Then I demonstrate how this effect operates in a powerful way within an important religious context like the mystical journey—specifically the mystical ascent of Saint John of the Cross and a few other mystics.

The Mountain Climber and the Prozac Effect

IMAGINE THAT AFTER OVERCOMING cancer a man decides to prove to himself that he is as healthy as ever by climbing Mount Everest. He trains his body, strengthens his mind, saves money, and makes all his plans. Then he goes and executes an extremely difficult (and expensive) climb. He makes it to the top, and though nearly wiped out by fatigue, as he stands on that freezing and windy summit he experiences a stunning elation. Never before has he felt so alive, so real. What kind of a feeling is this? Why does he have it? Could it be a lack of oxygen or his fatigue? Is it relief?

Before he consciously recognizes what he is feeling, his mind is already fast at work on giving the emotion some meaning—on making it a feeling.[8] He may not be aware of this process, which is an example of attribution. Few of us ever are. Now, had he failed to reach the top, he would probably have engaged in more intense and conscious attribution work in order to sort out the failure—we tend toward more intense introspection when we fail to achieve goals, but that is another matter. Here we are interested in success—or, to be even more specific, not success itself but the pleasure that accompanies the achievement.

According to researchers in this field of psychology, there are several (nearly instantaneous) steps that figure in the task of understanding the outcomes, motivations, and emotions associated with important tasks.[9] These steps constitute a sort of folk theory in the sense that the reasoning is rarely self-reflective or systematic. Still, there is a pattern here: the first step separates the *emotion* that attends the success of the task from the *causes* for the success (such as strength, training, or luck). The second step consists of uncovering the causes (this is "causal attribution") by following a set of criteria: internal versus external; fixed versus variable; specific versus general; controllable versus uncontrollable; intentional versus unintentional.[10] According to Thomas Duval, the work of attribution, the search for causes, is undertaken as tension builds between the perception of the external achievement and the internal self-evaluation.[11] If the achievement exceeds one's low self-esteem or fails to live up to a high sense of worth, a sort of cognitive dissonance emerges that needs to be resolved. The third step, Bernard Weiner argues, has to do with whether the outcome (reaching the mountain peak, feeling great) resulted from following a path that led to a goal with the help of conscious effort. After all, many outcomes—for instance, getting sick after eating dinner in a restaurant—have little to do with the path-goal relationship. The cancer survivor who reached the peak of Everest is more likely (but not certain) to identify the causes as internal (effort, character), relatively fixed, specific, somewhat controllable, and certainly intentional. He may thus attribute the strong affect on the mountain peak to these salient features and feel achievement pleasure (mastery), victory, or elation.

This picture is highly variable. If the climber has only just recovered from the illness and everyone, including himself, had grave doubts about his climbing Everest, then the outcome is so outstanding that the causes have to match it in magnitude. He may then attribute the result to an external factor beyond his control. It could be luck, the prayers of his loved ones, or God. We see this often with athletes who point to the sky or tell the interviewer that God made their win possible. In such a case, the pleasure would not be experienced simply as achievement pleasure, victory, or elation. The climber might attach his pleasure to the external factors and feel relief, gratitude, or perhaps even love. Finally, if the porters had to carry the climber on their backs, he might just think that he is enjoying the view, he might experience shame or guilt, or he might feel relief about having survived such a difficult test.

All of these are examples of the "Prozac effect" at work. The pleasure is *interpreted* in a particular manner only after the fact—it is hung on a peg, so to speak, that the mind builds for the pleasure with the help of complicated cognitive operations. None of these interpretations represent the true *cause* of the pleasure, which is always the adaptive function (novelty, mastery, exploration) utilizing the brain's machinery.

On his way to work in a government building, a man stops for a cappuccino with a double shot of espresso. As he drives down Interstate 270 toward D.C., the caffeine enters his bloodstream and he begins to feel lightheaded and euphoric. Suddenly, the traffic speeds up, everyone is smiling at him, and the day's humidity disappears. Well, not really, but it almost seems that way. His euphoria requires some explanation, and without really thinking about it, the man's mind provides the causes for the pleasant feeling. Peter Kramer, writing about the effects of Prozac, discusses this as the separability between brain states and observed mental objects. It is an error—the Prozac effect—to attribute the good feeling to whatever is present in consciousness when the brain (with or without the presence of serotonin reuptake inhibitors) manufactures pleasure.[12]

We see that the Prozac effect—the way we account for or actually experience a pleasure—is intimately connected with the causal attribution that explains our success. But it is important to emphasize that the feeling itself—for instance, elation—is already an interpretation of the pleasure based on a quick and nonconscious evaluation of the causes linked to the outcome. "Elation" is the Prozac effect term assigned to what may in fact be mastery pleasure or, beyond that, the playful and joyful exploratory pleasure that exists beyond both the novelty and its mastery. The same holds true for the love that religious people claim to feel as coming from God. Such an attribution error is only possible because what they truly feel is pleasure. Only deep pleasure triggers this type of effect. And because the claims of divine love by mystics who know God directly are the starkest and are generally regarded as the most authoritative, that is the topic to explore next. The following section outlines the mystical path as described by a number of prominent mystics. I show how the path produces profoundly hedonic states of consciousness, which must be interpreted by the mystic as a feeling of love coming from God.

The Mystical Path and Its Stations

In 1911 an extraordinary Englishwoman, Evelyn Underhill, wrote the first systematic modern study of mysticism (*Mysticism*), in which she defined the concept, in a neutral manner, as coming to know Reality directly and as it truly is. But what Underhill, an Anglican, meant by "Reality" was the world as God had created it, not the "blooming, buzzing and confusing" chaos that our senses perceive. A few years later, in an abridged version of her book titled *Practical Mysticism*, Underhill addressed what she called the "practical man" (all of us, I suppose) in a simple and straightforward manner.[13] She described the mystical journey every one of us ought to undertake if our life is to become meaningful. A distinguished scholar as well as a spiritual guide, Underhill identified three major steps in this journey: (1) beginning along the mystical path with daily meditation "constantly recapturing the vagrant attention"; then (2) practicing a balanced asceticism, a gentle renunciation along with unselfish service, that trains the will away from one's own concerns; and finally, (3) completely surrendering the self, an act accompanied by the gift of grace and illumination along with love—though periods of aridity and despair are to be expected.[14]

With this threefold scheme, Underhill provided a distilled version of numerous mystical accounts, which she had discussed in detail in her *Mysticism*. She was so committed to the notion of a mystical path that she even published a work entitled *The Mount of Purification*, drawing on the metaphor of the climb. This trajectory, more than the flat path, relies on the schema I discussed in the example of the Everest climber: a movement toward a highly prized goal at the expense of great effort, but with corresponding rewards. Indeed, the image of a mountainous climb toward God, or "Reality," is a virtual cliché in mystical literature, with Saint John of the Cross's *Ascent of Mount Carmel* holding the place of honor in the genre.[15] Underhill's map of the mystical path applies to the narrative of Saint John of the Cross (Juan de Yepes y Álvarez, 1542–91), even if the Spanish mystic's actual terrain is far more elaborate. A close reading of John of the Cross reveals the following key steps:

- Control of attention by meditation or prayer
- Discipline of the will by austerities
- Complete surrender or dissolution of the self
- Obtaining the rewards (joy, love, knowledge)

It is possible to divide all of this schematically into a beginning, middle, and final stage.

1. *Beginning:* The ascent begins with a stanza that the author interprets as his text unfolds:

One dark night
Fired with love's urgent longings
—Ah, the sheer grace!—
I went out unseen,
My house being now all still. . . .

The departure at night, according to the commentary, is a "privation and purgation of all sensible appetites for the external things of the world, the delights of the flesh, and the gratification of the will."[16] Clearly, for Saint John of the Cross the start of the mystical journey, the elementary techniques for beginners, require directing attention away from the external world and reining in the will. Mortification of the flesh, for example, is no mere masochism but a spiritual technique designed to contract the focus of attention from the restless senses and develop willpower to remain detached from distracting objects of awareness.[17]

2. *Middle:* After disposing of what John of the Cross calls "alien affections" and overcoming weakening appetites, the climbers are prepared for the next step—the changing of one's clothes, so to speak: "God, by means of the first two conditions, will substitute new vestments for the old."[18] This means that the individual self dissolves and the climber becomes identified with God, clothed in God's own clothing. This attainment requires, to stay with the same metaphor, a temporary but complete "nakedness," which is a death to one's biographical self, selflessness, or, in the language of the Church, abject poverty.[19] John of the Cross repeats, as do about half the mystics whom Underhill has studied, that the climber does not seek consolations or sweetness (rewards) but proximity to God, however painful that may be.

3. *End:* Finally one attains the reward of the climb—union with God. This comes about in a gradual manner, with what John calls "divine touches." But even those mere touches are so extraordinary that the devil himself could not have produced them as deceptive imitations. "These touches engender such a sweetness and intimate delight in the soul that one of them would more than compensate for all the trials suffered in life."[20] *The Ascent of Mount Carmel* does not provide a very detailed or lengthy description of the joyful rewards that await the climber at the top beyond such glimpses (the book is incomplete), but the peak—or rather, a number of peaks—has been described by Teresa of Ávila, the other Spanish mystic, who met John in 1567. Referring to one rapture in the next to last "dwelling place" of her masterpiece *The Interior Castle*, she wrote: "The joy makes a person so forgetful of self and all things that he doesn't advert to, nor can he speak of anything other than, the praises of God that proceed from his joy."[21] The joys of the final stages, which transcend everything that came before, are described in a wonderfully biblical metaphor: "Here one delights in God's tabernacle. Here the dove Noah sent out to see if the storm was over finds the olive branch as a sign of firm ground discovered amid the floods and tempests of this world."[22]

In sum, the main steps of the mystical climb can be described as *detachment*, *dissolution*, and *union* (or reward). These may resemble the way in which ordinary attention works as it guides us from one object to another: detachment from the first object, transition, then attachment to the new object. This may seem like a superficial observation; after all, the mystic detaches attention not from a single object but from the world as a whole. Furthermore, the three steps (detachment, dissolution, union) are already a gross oversimplification of a very elaborate journey that varies from one individual to the next and from one religious tradition to the others. Still, comparing the mystical path to the dynamics of attention-shifting makes more sense as we consider that virtually all the mystics emphasize the general pattern that Underhill has observed, even acknowledging the great variety of mystical accounts.

For example, the medieval German mystic Henry Suso (1295–1366) shared little historical or geographical proximity with John of the Cross, but his spiritual biography, *The Exemplar*, offers up a very similar pattern:

1. *Detachment:* This first step seems to be Suso's specialty—he gained fame for his extreme self-mortifications. But there was a system even in the disciplines, namely, turning attention away from external distractions. One example was a vow of silence while eating, a vow that could only be suspended at the permission (in a virtual sense) of Saint Dominic, Saint Arsenius, and Saint Bernard. Such permission was granted on the condition that the conversation involved neither external attachment nor internal disturbance.[23] This was also the purpose of Suso's meditations, such as the following extraordinary one:

> Dear Lord, on the high branch of the cross your bright eyes were darkened and rolled. Your divine ears were filled with the sounds of scorn and curses. Your refined sense of smell was defiled by vile stench. . . . And so I ask today that you keep my eyes from wanton looking about and my ears from listening to empty words. Lord, take from me the taste for material things.[24]

2. *Dissolution:* For Suso, dissolution means suffering, a subject about which he writes in great detail. Naturally, there are many different types of suffering, but the one that just precedes the great joy of spiritual accomplishment is unique. It is the surrender of one's very being, a form of suffering modeled after the example of Christ.[25] The cause of suffering may be physical hardships or, worse, the feeling that God has abandoned the seeker. But Suso reminds his readers that the seeker ought to see that through suffering God has made him equal even to the martyrs. This can become a source of great happiness as long as the suffering is accepted in the spirit of self-surrender.

3. *Reward:* Suso describes a generous mystical reward for the servant who truly suffers for Christ.[26] It is a stunning unity: "He was transported into himself and beyond himself. And in a state of withdrawal from his senses something spoke sweetly within him." What God told the servant on that occasion and on others was that those who "died with me in joy they shall also arise with me." They shall receive extraordinary gifts, such as heavenly rewards, divine peace, and, above all, "I shall kiss them so intimately and embrace them so lovingly that I am they and they are me, and we shall remain a single one for ever and ever."[27] This joy is not a future promise but shall begin immediately and be enjoyed eternally once the union is reached.

Because both are Christian, Saint John of the Cross and Henry Suso can be interpreted as heirs to a Christ-based theology in which the three steps of crucifixion-death-resurrection give shape to the mystical imagination. Would a Muslim mystic also emphasize the three steps of detachment, dissolution, and reward (union)? The answer is yes.

According to the eminent Islamic scholar Annemarie Schimmel, Islamic mystical experience consists of a path (*tariqa*) along which the seeker journeys, with several distinct stations. Schimmel quotes Yahya ibn Mu'adh of Iran, who names repentance, asceticism, peace in God's will, fear, longing, love, and gnosis.[28] Other Sufi sources give additional stations, but Schimmel extracts the universal ones: repentance, trust in God, poverty, contentment, love, and gnosis. Do these correspond in psychological terms to the Christian detachment, dissolution, and union? Instead of gathering illustrative quotes from several Muslim mystics, which can easily distort the evidence in my favor, I shall look at one example.

There have been more famous and more prolific mystics (none more than Rumi or Al-Ghazali), but among the women, Rabi'a al-Addawiyya (717–801) stands out as the most prominent. She was also an extraordinary and highly eccentric mystic, worthy of quoting in this context. Rabi'a, who lived in Basra in Iraq, did not leave her own written works, but Margaret Smith has assembled a significant number of references to her ideas. These testimonies were authored by such prominent sources as Al Ghazali, Abu al-Qasim al-Qushayri, Abu Nu'aym al-Isfahani, and others.[29] The assembled sources paint a vivid picture of an extraordinary woman, along with a highly respected, even revered, account of the major steps leading up the mystical path.

The first step is repentance (*tawba*), which Abu al-Husayn al-Nuri defines as "turning away from all save God Most High." Rabi'a, outdoing other Sufis, claimed that even repentance depends on God. "If I seek repentance of myself, I shall have need of repentance again."[30] In other words, the initial turn away from the world already implies a diminished ego, or dissolution. But it takes place more dramatically in the steps (poverty, renunciations, unification) that lead up to the recognition of God's unity (*tawhid*), a truth that annihilates all else, including the personal self. In the words of Abu Sa'id b. Abi al-Khayr, "until you empty yourself of Self, you will not be able to escape from it."[31] Rabi'a once responded to a question about conditions for reaching proximity to God: "That the servant should possess nothing in this world or the next, save Him."[32] The reward, if one can call it that, is the step called fellowship (*uns*), which is described by al-Junayd as the "abiding remembrance of God in the heart, in joy and delight and great longing for Him and intimacy with Him." At this peak of mystical experience, Rabi'a described two types of love, one "a selfish love," the other "a love that is worthy of Thee."[33] These were interpreted by Al-Ghazali in the following manner:

> She meant by the selfish love, the love of God for His favour and grace bestowed and for temporary happiness, and by the love worthy of Him the love of His Beauty which was revealed to her, and this is the higher of the two loves and the finer of them. The delight arising from the Beauty of the Lord is that which the Prophet of God explained when he said speaking of his Lord Most High, "I have prepared for my faithful servants what eye hath not seen nor ear heard and what has not entered into the heart of man," and some of these delights are given beforehand in this world to the one who has wholly purified his heart.[34]

The Hedonic Psychology of the Three Steps

In the previous section, we saw that the long journey of mystics can be divided into three major phases—detachment from the world, dissolution of self-identity, and achievement of the goal, such as union with God. Although this scheme greatly simplifies the true state of affairs, there is strong evidence to support it. Furthermore, even a schematic map cannot obscure the striking fact that the last step virtually always brings extraordinary levels of pleasure, described as joy, rapture, bliss, sweetness, ecstasy, and so forth. Thus, while on the level of faith or theology the final goal is knowledge of God or unity with Him, the secondary but universal effect is a feeling of pleasure. Why is this? Why is it that mystics from every tradition and every period in history have described intensely positive states of consciousness? Consider just the following words of the great German mystic Meister Eckhart (ca. 1260–1327):

> In this agent, God is perpetually verdant and flowering with all the joy and glory that is in Him. Here joy so hearty, such inconceivably great joy that no one can ever fully tell it, for in this agent the eternal Father is ceaselessly begetting His eternal Son and the agent is parturient with God's offspring and is itself the Son, by the Father's unique power.[35]

If we take the first two steps of the mystical journey, which may account for the third (pleasure), and analyze them psychologically, we see that pleasure in its elevated forms is a natural outcome of the mystical path—even if the mystic does not believe in God or in divine rewards. The key is attention-control and the dynamics of self-identity.[36]

Mystics are specialists of attention, but they are not the only ones: in a recent tournament, Tiger Woods demonstrated his extraordinary capacity for tuning out the crowds, among which he has to perform, in order to make a difficult putt. Walking down the fairway, he kept his eyes on the grass in front, or on a distant horizon, while his playing partner smiled and waved to the cheering audience. You can guess who won. The next day the *New York Times* columnist David Brooks marveled at Woods's level of concentration and speculated about his Buddhist mother and disciplinarian father. Indeed, the awe that this golf player's mental powers generate borders on the mystical. Alas, my cat can focus just as well, and for longer durations, when she knows where a chipmunk has made its little home. But no one would consider the furry predator a mystic.

So what is this extraordinary attention? Psychologists define attention as "a limited capacity mechanism that mediates conscious awareness of internal and external events."[37] But this does not really help. What does "mechanism" mean in this context, and how does it "mediate"? Is it the same as awareness or consciousness or a controller at the switch of a film projector? And why would the tweaking of attention (by mystics) lead to pleasure? Surprisingly, researchers still do not have a clear idea about the nature of attention.[38] However, it is not difficult to describe. Imagine that a friend brings you along to a Beverly Hills party hosted by Tom Cruise. As you casually mingle with the guests, two things suddenly happen at the same time: a waiter stops in front of you with a tray of hors d'oeuvres just as Tom Cruise reaches out to shake your hand and introduce himself. Which do you notice and respond to? Which will you later remember? Did you voluntarily ignore the waiter, or did it just somehow "happen" that you noticed the famous host?

This is an example of attention performing one of its two major tasks: selecting one among two or more competing items.

Imagine that just then the waiter dropped his tray on your rented tuxedo. Would you keep your eyes focused on Mr. Cruise? Probably not—your attention would somehow shift quickly to the accident, which you would consider a distraction. There is only so much we can perceive at one time—or to put it technically, attention is subject to capacity limitation. When the load threshold is breached—shaking a legend's hand as oysters go flying onto your trousers—attention shifts to the second event. That is why we call it a distraction.

All of this is fairly obvious. Early psychologists, like William James and Edward Titchner, who worked on this subject with the help of systematic introspection, have provided us with a commonsensical understanding of what attention does.[39] It guides awareness, subject more or less to the intention of the subject, like the hand that holds a flashlight in a dark room. The greater the mastery of will or the power of conscious effort, the easier it becomes to disengage from one object, move to the next, and keep it there. Since the middle of the twentieth century, researchers have not been so sure; some cognitive psychologists today do not think that the concept of attention even means anything.[40] All that truly exists, they argue, is computation, or the processing of information where the machinery of the brain—like a computer—can only compute according to given rules and limited capacity.

Biological theories of attention are more interesting, in my opinion, than cognitive models, and they probably account for how attention-control causes pleasurable feelings. Whereas a computer is programmed to attend to specific information, the organism—including humans—responds to salient information in the environment (external and internal). In other words, some features of the world are threatening, and others are attractive for eating or for mating. Adaptation to natural environments and the propagation of the gene require that we be able to tell one from the other, select the stimulation that is most pertinent at a given moment (while excluding others), and act appropriately.[41] This needs to be done quickly, that is, without the intervention of higher reasoning or a self-aware subject. Salient events, and virtually all new ones, generate emotion and affect—both positive and negative—almost instantly, and these guide attention in an adaptive manner. Imagine that while jogging in a stadium you notice other joggers moving in the same direction as yours, but then one runner suddenly cuts across the field and sprints in your direction. Your heart begins to race and you brace yourself. If one of the other runners were just then to fall down with cramps, you probably would not notice at all. Your fear would keep you focused on this one threatening runner.

The brain mechanisms that have evolved to "control" our attention—actually, it is not control but appropriate response—are reasonably familiar to neuropsychologists. Organically speaking, the attention system (thalamic nuclei, anterior cingulated gyrus, prefrontal cortex) is embedded with limbic projections (from the amygdala, hippocampus, and orbital frontal regions). This accounts for the fact that attention is connected to motivation and reward systems, which makes sense if paying attention to salient information allows us to make the beneficial (pleasurable) choice. The entire systemic function of attention depends on neuromodulators and chemicals such as norepinephrine, serotonin, dopamine, acetylcholine, and beta-endorphin. These help produce, not a single diffuse state of motivation (which is necessary for arousing attention), but a number of distinct and different-feeling motivational states (both wanting and liking).[42]

Now, imagine returning to the same stadium where you thought you were about to be assaulted. If that other jogger is there, you will remember him well, but none of the other runners. You feel a bit tense around him and keep an eye on him. Technically speaking, your attention has been "primed" to notice him and keep him in focus.[43] The priming or preparation of attention is enormously powerful—it literally makes us perceive some things in the world and completely ignore others. Priming may actually make us blind to events that are taking place right in front of our eyes. This phenomenon, called "attentional blindness," has been demonstrated in a number of famous experiments. In one of them, students were instructed to view a video showing two teams of players passing basketballs back and forth in a closed room. They were instructed to count the number of passes made by one team and thus failed to notice a man wearing a gorilla suit in the very center of the room.[44]

This is a very dramatic example; most other studies have focused on the more subtle influence of priming on the ways in which subjects perceive sounds, colors, words, and other types of information. But the conclusion is widely shared: it is easy to direct attention by cuing emotions and expectations, by giving a variety of distractive instructions, by setting goals, and by using deception and a variety of other techniques.

Mystics have their own attention laboratories, and they too work with powerful priming techniques. The interesting and unique feature of mystical priming is its strong hedonic component. On the surface level, these mystical techniques have only the appearance of ethics, but they work as attention regulators thanks to their emotional and affective power. For example, if a novice learns from her master that food craving is a sin, she will find it easier to disregard, even become blind to, signals of hunger. If a monk learns early on that music is a frivolous waste of time, he will find it easier to tune out sounds coming from outside his cell. It may appear that the ethical training of the first step of the mystical journey is only the strengthening of moral resolve—the will. That is how the texts put it. But rules about sexual behavior, eating, music and dance, gossip, daydreaming, or material goods also define what constitutes salient information. If you cannot master these ideas ("temptations") early in training, you will not be able to control your attention later on, despite heroic efforts. If you do not come to truly believe that the world is either evil (disgusting) or an illusion, you will not be able to maintain focus away from the world. These beliefs require an institution, a church, a monastic order, or a teacher to drive home the point early in the mystical training.

Once attention-control is mastered, tuning out the world becomes easier, second nature. And the more effective the shutdown of external and internal stimulation, the further away does the self dissolve as ever more exquisite pleasures arise. There are organic reasons for this. According to Andrew Newberg's summary in 2005 of neurological studies of mysticism, a number of brain systems and several neurochemicals are implicated in the dynamics of the mystic's shutting down of distractions, deepening and emptying of focus, dissolving sense of self, and deep euphoria.[45] The organs include the prefrontal cortex (PFC, right and left), the cingulate gyrus, the thalamus (and thalamic reticular nucleus), the posterior superior parietal lobule (PSPL), and others. These operate in a number of systemic loops of feedback and feed-forward toward the achievement of something Newberg calls deafferentation (disconnection of nerve signals), with its very powerful consequences (which I have called dissolution). To give one example, the practice that allows the voluntary shutdown of distractions is actually the increased activity in the thalamic reticular nucleus. It is as though it was some internal "white noise" that kept you from seeing Tom Cruise rather than the falling platter. The thalamus is a major relay station for incoming sensory, emotional, and physiological information, which is controlled (by the cingulate gyrus and right PFC) or completely eliminated by the increased level of activity in those centers. The higher the activation of one center, the more effectively it shuts down the information in the other, including information on which the individual organism depends for its perception of the self.[46] At the same time, a variety of pleasures emerge from the mystic's power to control the attention. For example, controlling your breath as you do in meditation triggers the production of beta-endorphins, causing joy and euphoria. In contrast, the voluntary decrease in sensory input, correlating with the increased activity in the visual centers (internal imagery), accounts for a rise in serotonin, which results in a more moderate mood enhancement.

In sum, we see here that the first two steps of the mystical path (detachment from the world and dissolution of self) converge in the psychology and physiology of the brain. Attention control, at its highest levels, not only shuts out the world but actually increases the activity in brain systems that block, or gate, the functions of self and at the same time produce a number of distinct types of pleasure. What is the nature of these pleasures from the perspective of evolutionary adaptation? Recall that I have discussed three distinct types of adaptive pleasure ("type" refers to the true cause of a particular pleasure, not its feel). The first and simplest is novelty pleasure, what we normally call sensory pleasure. The second is mastery pleasure, which is felt as empowerment and achievement. The third is exploratory: pleasure that feels like joy or euphoria and takes for granted—moves beyond—the fulfillment of needs or the complete mastery over them. The first step of the mystical journey controls one's enslavement to novelty pleasure, the hedonic treadmill. It brings about a sense of external mastery. The second step, the diminishment of the self, produces the satisfaction feeling of internal mastery. But these are relatively low levels of enjoyment. The dominant mystical pleasure at the height of extreme attention control is of the third type—joy, bliss, rapture, or ecstasy. It is not a need fulfilled or the ego's mastery over its desires. It feels like an unearned gift, a spontaneous event, a blessing that comes out of the blue or without the self's involvement. Here is how the Indian mystic Jnaneshvara (died c. 1293) described it: "The mind became immediately composed. Internally there was a feeling of joy. . . . As an unblown lotus swings to and fro on the surface of water on account of the bee which is enclosed within its petals, similarly, the body of the devotee began to shake on account of the feelings of internal bliss."[47]

Such a pleasure does not reflect on the self's achievement because it emerges spontaneously; furthermore, there is no self, or the self is much diminished. And because this pleasure, like the others, is actually a functional and chemical brain state, it feels "vertical."[48] This means that one cannot properly describe it or even name it in any meaningful way. Despite its saturating presence, this pleasure lacks the sort of texture that would allow the mind to sink its teeth into it. It can no more be described than the pleasure of cold ice cream on a hot summer night. It is just a state of intense and diffuse "goodness." That is when the attribution process kicks in.

Like the climber on top of the mountain, the mystic experiences an intense affect that has to be named or somehow interpreted by the brain. I believe that those mystics who speak of receiving God's love at the peak of their journey are interpreting their pleasure in such a way. They are exhibiting an extreme (and rare) case of the Prozac effect—a hedonic attribution error. Note that I am referring to receiving love, not offering it up—as many religious people do, even nonmystics. But why love?

The Origin of God's Love

DESPITE WESTERN CULTURE'S LONG history of love and lovers—think of Paris, Romeo, Orpheus, Tristan, Abelard, and all the rest—no greater lovers have existed in that culture than mystics who love God. No one has loved more deeply or against greater odds than Saint Augustine, Catherine of Siena, Julian of Norwich, Rumi, Rabi'a al-Adawiyya, and all the other love mystics. Consider the words of Augustine:

> O Eternal Truth and True Love and Beloved Eternity! You are my God, to you I sigh day and night. . . . And you beat back the weakness of my gaze, powerfully blazing into me, and I trembled with love and dread.[49]

These mystics were motivated to seek God by intense love, which helped them summon the energy to focus attention and discipline a feeble will to the point where they could later describe the ultimate reward: God's overwhelming love in return. Where did that love truly come from? Must we assume that either God exists (and flooded them with love) or all of these mystics were "delusional" (in the language of Richard Dawkins)? I shall assume that their reports were accurate—that they were not imagining the love about which they wrote. At the same time I shall also leave God (whatever that term means) out of the explanation for the feeling they reported. Keeping things natural, I shall argue that the love that mystics feel radiating from God and entering their very being is the product of the Prozac effect: it is most likely a case of attribution to an intense joy, interpreted in light of the mystics' own feelings of love for God.

The key to the love of God—and indirectly to the theory that God *is* love—is the verticality (or invisibility) of pleasure and the mystic's own feelings of love. What kind of love is that? What, indeed, is love? Does the mystic love like a parent, a friend, a lover? Contemporary psychologists divide love according to a number of taxonomies in order to explain its different social and biological functions, while accounting for the distinct neurological pathways that underlie each function. Ellen Berscheid, for example, distinguishes within the broad category of love the following subcategories: attachment, compassionate, companionate, romantic, and lust.[50] The first, for example, is evolution's way of guaranteeing an innate attachment to a caregiver who is essential for survival. The second, which is Abraham Maslow's famous "B-love," is also an evolutionary product geared toward caregiving and social support. According to Clyde Hendrick and Susan S. Hendrick, romantic love itself can be divided into a number of "styles."[51] This taxonomy includes *eros* (physical attraction), *ludus* (love as game), *storge* (friendship), *pragma* (matchmaking for useful qualities), *mania* (unhealthy mood swings), and *agape* (self-sacrificing love).

Despite these eccentric-sounding names, most contemporary psychologists of love agree that the categories or styles of love transcend cultural differences. Granted that different child-rearing practices along with varying ecologies and economies cause distinct cultural approaches to love, it is still possible to speak of universal and biological foundations. It seems then that both evolutionary and neurological theories of love represent a sensible approach to explaining the love that a mystic may feel toward the "desired object."

Although some mystical writings give the appearance of erotic or lust-driven love, I shall assume that this is a literary device. Furthermore, although leading psychologists, beginning with Freud's nineteenth-century mentor Jean-Martin Charcot, have not shied away from exposing the sexual underpinning of mystical desire, I shall stay away from this singular reduction.

It is safe to describe the love of the mystic as both attraction and attachment. By "attraction" I refer to Berscheid's "romantic" love. This does not require the physical presence of a beloved, or even that the beloved be embodied. The essential characteristic is strong motivation toward the other, which energizes the lover, enables focused attention on the object of love, and in extreme cases can lead to obsession (exclusive and constant attention) with the object of love. Consider these love words of Rabi'a al Adawiyya:

O my Hope and my Rest and my Delight
The heart can love none other but Thee.[52]

According to Helen Fisher, these feelings are marked by elevated activity of dopamine in the reward pathways in the brain along with norepinephrine, which mediates attention and allows it to exclude distractions.[53] This seems to confirm the work of researchers like Kent Berridge and his colleagues, who have separated wanting from liking and who argue that dopamine accounts for the former.[54] At the proper brain center, it is dopamine-mediated states of wanting or heightened motivation that make a subject give up food and sex for heroin and lead to prolonged and intense focus of attention on the desired object (the fix).

The mystic's second type of love is attachment, or companionate. Unlike the passionate and active energy of attraction, attachment love produces states of calm, security, social comfort, and emotional union.[55] The underlying brain chemicals are probably oxytocin and vasopressin. Psychologists know that both attraction and attachment love are associated with biological reproduction and with social bonding. This does not mean that the mystic who feels love for God is acting, like a college sophomore, according to a strong biological impulse. Nor, again, does this mean that God is a delusion. It only means that even the loftiest religious feelings and expressions are served by the natural hardware that has evolved in response to social pressures. But this perfectly natural love felt toward God helps to explain why mystics report being loved in turn by God. That reciprocating love, which is felt as supernatural and divine grace, is the reward for the love of the mystic. The mystic feels that her Beloved loves her in return. But how can this be? If we reject supernatural explanations and yet accept the reports of mystics as accurate, we must explain how an idea (God) can possibly love back.

God's Love as Prozac Effect

It is time to put together what we know about mysticism, pleasure, and love. Advanced mystical discipline almost always produces intensely positive (pleasurable) states of mind—joy, bliss, rapture, and so forth. Pleasure, as we have seen throughout this book, is virtually impossible to describe; a body-feeling without contours, it is what I have called "vertical." Even the cold, sweet, or fatty qualities of ice cream are not the same as the pleasure of eating it. But the pleasures of mystical experience are extraordinary—intense beyond anything else the mystic has ever experienced. For the mystic himself, the pleasure has no obvious cause: there is no external object to associate with the pleasure (like ice cream), and there is no obvious internal trigger, such as an ego, that caused it (like the winner of a marathon). As a result, the pleasure feels spontaneous.

Attribution theory posits that an outcome—and pleasure is a very powerful outcome of the mystical path—must be accounted for both as the effect of some cause and as a feeling (an interpreted emotion). As an effect, strong pleasure can only have been caused by a powerful agent proportional to it in magnitude. And because the pleasure feels spontaneous, this agent must be external to the mystic himself. Furthermore, because the peak pleasure emerges at the end of a goal-directed path, and the goal was God, the pleasure can only be a gift or a reward from God. It does not matter whether the mystic is a Spanish Christian, a Polish Jew, a Baghdad Muslim, or a Bengali Hindu.

The feeling component is a bit more complex. We have discussed two modes of love toward God, attraction and attachment. Both of these also include a dimension that I have not discussed—sacrificing love, which is often discussed in connection with agape.[56] Both modes of love, and the sacrificing dimension as well, have a strong hedonic tone. Attraction operates on dopaminergic principles similar to those of other great motivators—the desire for food, gambling addiction, the mesmerizing power of new products at the mall. Technically, these are examples of wanting, not liking, but because they act like magnets on us, we experience them as pleasurable. Attachment love feels more satisfying than attraction, like resting in the shade with a glass of cold water. Sacrificing love provides an ample feel of mastery pleasure (over one's own needs), while ludic love is playful. Only a careful study of a specific mystic's autobiography or diary can reveal which type of love he or she is feeling toward God throughout the journey; furthermore, love can also hurt if the object of love fails to respond or reciprocate. But love fulfilled is intensely satisfying.

The pleasure felt as reward from God can only be interpreted as love. Keep in mind that being loved is not a feeling, properly speaking, but a judgment, an appraisal of signals received from another agent. In the absence of the Lover Himself (I know I am taking a metaphysical position here), some hypothesis must be generated. And as folk theory, nothing works better than a proposed feeling that is thoroughly hedonic and powerful and corresponds with the same feeling the mystic himself brought to the journey. The mystic who does not believe in God or for whom the final experience is not the effect of some cause—the Advaita Vedantist or the Rinzai Zen Buddhist—need not interpret the joy as love. For the Hindu, as one example, the pleasure (*ananda*) is simply a characteristic of ultimate reality (Brahman) along with two others: *sat* (being) and *cit* (consciousness).

In conclusion, if my analysis of God's love is correct, the main consequence is cautionary: it is impossible to construct a mystical theology (a theory of God's existence and qualities) based on the experience of God's love. This is not a huge loss to theology; theologians will find other grounds for defining God as love. Furthermore, the idea of being loved, though based on an erroneous folk theory (attribution), is not itself a bad thing—when it comes to mystics. After all, mystics represent a tiny minority of religious practitioners, and they are rarely individually influential. However, the idea of God's love has entered the mainstream of the three Western monotheisms. For example, God's refracted love can act as the basis of religious ethics or at least as a motive for loving others. One example of where this may lead is probably not atypical:

> Dear God, Some of your children are extremely irritating and, honestly, difficult to love. I don't really want to be around these people, but know that I am called to reflect your love to them. This is really gonna need to come from you.[57]

In other words, God is called on not only to reward the lifelong mystic but also to enable ordinary human encounter. But we have to wait for Him to get involved.

For the majority of religious people, divine love manifests not as a direct experience of God but as a way of interacting with some human intermediary. It is not God who loves, but the rabbi, the priest, the shaykh, the tzadik, the guru. This too is not necessarily a bad thing, unless love comes wrapped in a demand for obedience, which, in turn, is subject to gaming. That is the subject of the next chapter.

CHAPTER 7

Spiritual Love and the Seeds of Annihilation

THE NUN IN HER CLOISTER may develop a loving relationship with God, and the mystic in the mountains may live his whole life consumed by such love. The few individuals who attain this rarefied experience seldom move around much—their lives turn in narrow circles around this vertical relationship.

But Leviticus 19:18 made it a commandment that we should love our neighbor, and Rabbi Akiva, the second-century Jewish authority who was martyred by the Romans, declared this love to be the very center of the Torah (Sanhedrin 38a). Jesus, as reported in John 13:34–35, also thought as much: "A new command I give you: Love one another. As I have loved you, so you must love one another. By this all men will know that you are my disciples, if you love one another." Finally, back to Judaism and modern times, Martin Buber argued that love is the relation between two subjects, an I and a Thou.[1] It is the relationship that mirrors the love of God, who is the eternal Thou. This familiar idea is simply the tipping over of a vertical relationship, the love for God and by God, onto human relationships in the world. If one can love God, one surely ought to love another human. In practice, however, no such horizontal love can be detected in large-scale groups.

The majority of humans, even those who believe in God, mostly just love their relatives. Social groups hold together not because of spiritual love but as a matter of balancing competition for resources with a minimal civic responsibility. The better individuals live by honor, perhaps obedience, occasionally even charity. But love? Divine love? The perfectly vertical does not translate too well to horizontal relationships, not even given the injunctions of Jesus in the New Testament or the beautiful words of a Martin Buber.

The incommensurability of divine and human love is so stark that one wonders if every effort to reconcile them could only result in violence. Consider the relationship between Abraham and the Hebrew God (Elohim, El Shadai, and eventually Hashem) in the book of Genesis (22:1–24). When God finally asks Abraham to sacrifice his beloved, his one son, Isaac, in order to demonstrate by means of blind obedience his own love of God—the text is clear about this being a test—Abraham immediately agrees. And when the angel of God holds back the father's raised arm and a substitute victim, a ram, appears in the bush, how thrilled is Abraham? That is hard to answer because the text is mum. But the great medieval commentator Rabbi Shlomo Yitzhaki or Rashi (1040–1105) explains that Abraham begged God to let him, at the very least, cut the boy and inflict some injury.[2] If this test of love is a sort of initiation, just as Abraham's circumcision was earlier, should there not be some demonstration of pain? Is pain not the central moment in a proper initiation, even when the goal is love?

For many, this biblical narrative is a testament to simple obedience to the will of God, but to Rashi it was clearly about love. Begging for at least a cut reveals the passion beneath Abraham's obedience, the impulse that the great French psychoanalyst Jacques Lacan regarded as desire.[3] Can this enthusiastic obedience to Law (God's commandment or, barring that, any source of authority) contain the vertical love of God and extend it to the social realm? I'm not sure what the answer to that question may be, but I am reminded of an amusing anecdote from the Israeli press. A few years ago, the Israeli parliament passed a law requiring every driver to carry a reflective vest in his car, in case a flat or some mechanical problem occurred that required the driver to step outside the car and stand on the side of the road. It was a simple safety law, but thousands of drivers went beyond conformity: instead of keeping the light vest in the glove compartment or trunk, they slung it over the driver's seat. It became a sort of fashion statement, a "look." It seemed to say, not only do I obey the law, but I love to do so. The vest and the law are now a part of who I am, just like the car's paint job and the *hamsa* (good luck charm) pendant hanging from the front mirror.[4]

I know this is mildly amusing and in no way resolves the tension of extending vertical love outward. The Israeli driver may be acting like Abraham writ small in relationship to the law—but the gap between joyful conformity to the law and love remains unbridged. In fact, this only sharpens the original question: what kind of a community could ever revolve around vertical love and be held together by a feeling rather than mutual obligations or negotiated compromises? Clearly this cannot be a very large group. What first comes to mind as potential candidates are communes, cults, small monastic clusters, or, looking further back in time, the Jewish Essene group, the early Christian communities, and medieval mendicant groups. But we need to be careful here. The great German sociologist Max Weber wrote a very influential book on social groups that organize around charismatic leaders.[5] If at one end of this spectrum is the group held together by the force of a leader's personality acting as a magnet, in time this force can become institutionalized. Other individuals begin to assume specific governing roles in the community, rules emerge, and the early charisma gives way to bureaucracy. The types of groups I have listed here are not necessarily charismatic. To give one example, monastic orders, even when led by an extraordinary figure (for example, Saint Francis of Assisi), still develop the organizational rules that call for obedience, the vows that enforce the rules, and even the physical walls of the (Franciscan) monastery. That is not exactly what Weber regarded as charisma, let alone what I would consider love.

In some ways the "love community" is far more radical than even the secretive early Christian groups, about whose internal problems Paul wrote repeatedly and with concern. Such a group can have little in the way of rules, institutions, or enforced boundaries. The group coheres only by means of a bond that refracts the vertical love (sometimes called "friendship" or "agape") of a single mystic radiating outward to his or her immediate followers. Such groups are not common, but I am personally familiar with a few. Swami Vidyadhishananda Giri, a Hindu Kriya Yoga master and spiritual teacher who lives in the United States, commands far more than the respect and attention of a broad swath of American Hindus.[6] He also attracts a smaller group of immediate disciples and devotees who are drawn to his company by the spiritual mastery and love that radiates from him in a truly extraordinary manner.

Examples of such informal communities abound in all the major traditions: Sufi disciples who cluster around an extraordinary shaykh (master) or *pir*, Hasidic followers of a tzadik, Hindu devotees of a guru, Christian loyalists committed to a heretic teacher, such as Jon Huss or George Fox, who may or may not disperse after the teacher's martyrdom and/or imprisonment. These groups are fluid—members come and go, clustering around a single figure, then disbanding or becoming institutionalized and losing the initial emotional impetus. They are usually small but can attain great size and influence, such as in the case of someone like Sabbetai Tzvi, the Jewish false messiah. Some of the groups are marginal and passive, and others attain renown and political influence—such as the followers of Mohandas K. Gandhi and Martin Luther King Jr.—at least for a short duration. The groups I am discussing still exist within the great religious traditions and in countries such as Iran, Pakistan, Egypt, Israel, and India.

Despite their great diversity, some general features apply across the board: The greater the spiritual mastery of the leader, the greater is the love that holds the group together around him. Regardless of whether the spiritual master is Muslim, Jewish, Hindu, or Christian, the more passionate this divine love that flows within the group, the more dangerous this group may become under some circumstances, especially in today's volatile world. Again, it is not so much what these groups believe about God or the afterlife that threatens extreme violence, but the love and the social arrangement based on love.

This chapter focuses on these groups. It describes the ways in which divine love translates into a unique form of social clustering. Focusing on the master, the disciple, the initiation, the austerities, the erotic language, and the discourse against the center and political engagement, this chapter gives examples of groups from four religious traditions—Islam, Judaism, Hinduism, and Christianity—and explains what makes them both sublime and dangerous. A number of such groups in the Middle East are cited, and I conclude with a list of warning signs to look for as the Middle East begins to acquire nuclear arms.

The Spiritual Master

THE HISTORY OF RELIGION is rich with religious leaders of every conceivable kind, from prophets, mystics, and saints to heretics and scoundrels. Many of these have been members of institutions, such as the Catholic Church (popes, bishops), mainstream Judaism (rabbis), normative Islam (qadis, mullahs), and Hindu pundits. More unusual is the spiritual master, a man or woman who may or may not be part of an institutional order. The spiritual master is a leader of a limited, often tiny, group of followers only by virtue of direct spiritual experience and the personal charisma that such an experience generates. Looking at world scriptures suggests that "founders" of new religions (Jesus, Muhammad, Buddha) may have been such teachers. But there have been dozens of prominent examples and hundreds of lesser cases of religious groups that cluster around a spiritual master. Such masters are often barely known and frequently depart the world with barely a ripple, leaving little in the way of an organization. There are exceptions, of course, and the two most prominent exceptions may also be the two most famous non-Christian spiritual masters in the West: the Hasidic Baal Shem Tov and the Sufi Rumi, whose own master, Shams al-Din Tabrizi, was equally notable.

Because the Baal Shem Tov (Besht), "Master of the Good Name," Yisrael ben Eliezer (1698–1760), was one of the most influential religious leaders from the exilic period of Judaism, it is hard to consider him anything but mainstream. His life and work reveal, however, the tension between charismatic spiritual leadership and institutional power. The Besht was born and died in poverty in rural Poland and worked hard to conceal his level of mystical insight. Although he is generally regarded as the founder of Hasidism, most experts today know that he encountered existing groups of Hasidim, whom he joined (in Podolia and Kotov). His magnetism emerged from three sources: an uncanny ability to diagnose mental and physical ailments, along with an intimate knowledge of folk remedies; profound mystical (Kabbalistic) experiences; and a joyful and optimistic personality. All of these led to his growing reputation as a miracle worker and drew an ever-increasing circle of passionate followers from both mainstream (rabbinic) Judaism and the less educated.

In 1746 the Besht wrote a letter to his brother-in-law, R. Gershon, in Jerusalem, a letter that reveals one of his mystical experiences:

> I climbed one step after another until I entered the messianic hall where the Messiah studies Torah along with all the Tannaim and Tzadikim and there I witnessed great joy . . . although I still do not know what caused it. And I asked the Messiah, when will you arrive among us? And he answered, You will recognize that time in the following way: when your teachings become known throughout the world along with all that you have achieved and everyone will attain the same heights as yours, that will be the time of salvation. I was sorry about the delay and wondered when the time will finally come.[7]

Despite his direct experiences of the Shekinha, the Divine Immanence, the Besht recognized that his work toward salvation involved the repair of human sin and that he could do this work only by lowering himself to the humblest levels of human existence. His travels in the villages and small towns of Podolia, the miracle working, healing, and exorcizing, were all part of his saving work in "hell." However, his immensely joyful demeanor, his notions of *hitlahavut* (spiritual enthusiasm), and his *devekut* (passionate attachment to God) proved magnetic.

Among his many prominent followers, two became most influential: the Maggid of Mezeritch and R. Yaakov Yosef of Polana. Each exercised a different style of leadership: The first instituted the concept of *hatzer*, the court of the tzadik—a stable place of pilgrimage for followers. The second maintained the master's original style of wandering the countryside, visiting lay followers, and instructing them in people's kitchens and dining rooms. Each in his own way gradually changed what had been known as *havurah* (informal group) into a more formal institutional setting. In Max Weber's terms, the original charisma was diminished into a bureaucracy. This is not to say, however, that individual Hasidic leaders (tzadikim, rebbes) would not later, in their own way, draw a passionate following by the force of their own charisma.

For every renowned teacher like Besht there have been hundreds of largely unknown masters who were equally beloved by their direct disciples.[8] At times it is the disciple who makes the teacher famous, and no disciple in any tradition has been more widely read than Jalal al-Din Rumi. It is largely thanks to Rumi that we have learned about Shams al-Din Tabrizi, a Sufi master who might otherwise have remained known only to historians. Shams was born around 1180 in Tabriz, Iran.[9] Tabriz during the twelfth and thirteenth centuries was a major Sufi center with dozens of charismatic spiritual leaders, men who prized direct spiritual experience and love of God above theoretical knowledge or conventional authority. Many of these masters, like Khwaje 'Abd al-Rahim Azhabadi (d. 1257), are described as lacking all learning but miraculously displaying eloquence and insight in Qur'anic Arabic.

Shams was a spiritually gifted child with unusual proclivities, such as a penchant for fasting and stubbornly pursuing spiritual mentors. He spent a good bit of his time with men such as Shaykh

Abu Bakr Sallebaf of Tabriz, who taught him a variety of ecstatic techniques. However, Shams also acquired a mainstream theological education. He was tutored in the Sunni Shafe'i legal tradition and also became knowledgeable in alchemy, astronomy, logic, theology, and philosophy. Still, Shams took pains to conceal his erudition in simple company. He made his living modestly by tutoring and teaching in primary school.

Shams traveled widely in pursuit of a true spiritual master. According to his biography (*Maqalat*), "I implored God to allow me to mix with and be a companion with his friends. I had a dream and was told 'We will make you a companion of a saint.' "[10] He traveled to Baghdad, Damascus, Aleppo, Kayseri, Aqsaray, Sivas, Erzerum, and Erzincan. But in the final analysis, Shams considered himself an Ovaysi Sufi, one who achieves illumination directly from God, not through a shaykh: "Everyone talks of his own shaykh. In a dream the Prophet, peace upon him, gave me a ceremonial cloak (*kherqe*), not the kind that will wear out and rip after a few days and fall in the bath house and be washed of dirt, but a cloak of converse (*sohbat*), not a converse that can be comprehended, but a converse that is not of yesterday, today or tomorrow."[11]

When Shams and Rumi met, both were already well known. The meeting has been described in a number of versions as an extraordinary mutual revelation, a moment of perfect platonic love. Rumi, already spiritually accomplished, was not an ordinary disciple, such as a young initiate. Still, to the spiritual illumination of Shams's sun he was merely the moon, one who reflects that light to the common people. Rumi regarded Shams as the "Lord of the lord of the lords of truth." He wrote the following extraordinary lines about the man he regarded as his master:

The lady of my thoughts gives constant birth
She's pregnant but with the light of your glory
Shams al-Haqq of Tabriz, my heart's pregnant with you
When will I see a child born by your fortune.[12]

He also wrote the following:

Before the candle
That lights the soul
The heart
Is like a moth
In the flickering
Flame
Of the beloved the heart
Makes itself a home
A towering figure
Lion-taming
Drunk on love
A revolution
In the beloved's presence.[13]

Despite the obvious differences, the Sufi master, the Hasidic Tzadik, and the devotional guru hold a similar position in their small group as the magnet that keeps the band of followers together. For example, they all serve as explicit conduits to the love and insight of the divine. They are a sort of spiritual and affective *axis mundi*, the center of the cosmos. Ibn 'Arabi and 'Abd al-Karim al-Jili (1366–1403) describe the accomplished shaykh as a *qotb*, an axis or perfected man who mediates between the world and the energies flowing from above.[14]

The Hasidim use the image of verticality in the ladder of Jacob or the raised arms of Moses during the battle against Amalek. As R. Yaakov Yosef of Polana explained: "They constantly contemplated ways to ascend higher, and as a result the learned men, the zaddikim of the generation, became important in their eyes."[15]

In India, according to a medieval devotional text called Narada Bhakti Sutras, one attains the supreme bliss of unity with the universal Self through the grace of the guru—the guru mediates the love of God. Because the supreme object of love has no qualities at all, it must be mediated by love with form (*gaunabhakti*), which is embodied in the person of the guru.[16]

In the Caitanya Caritamrita, the great biography of the devotional Bengali Vaishnava leader Caityana (1486–1533), the guru is in fact God, and it is through the guru that God spreads his grace and love (1.1.28). That love is highly erotic, but the text carefully distinguishes between love as *prema* and love as *kama*, which is carnal and lower: "The *prema* of the gopis [shepherd girls] is the nourishment of the sweetness of Krsna; the sweetness grows and *prema* is fully satisfied. In the bliss of the object of love is the bliss of the container of love; bliss is not because of the wish for one's own happiness [but that of the object]" (1.4.168–69).[17]

The grace coming from the master, or from God but mediated by the master, can become known intellectually in the form of extraordinary powers. The Sufi shaykh possesses a power known as *baraka*, which is an ability to perform miracles or give blessings. The same holds true for the truly accomplished Hindu master, whose power, which may be known as *siddhi*, ranges from extraordinary healing to producing gold out of thin air—as the Sathya Sai Baba was wont to do. The Hasidic tzadik was also known for supernatural powers, as the Maggid of Mezeritch (one of the Besht's two major heirs) made famous. He is able to change shape or disappear altogether. The Besht himself was famous for his miraculous healing and diagnostics.[18] Despite these external powers, it is safe to say that the chief quality of the true spiritual master is the love that radiates from him and the joy that he gives his disciples by means of that love.

The Disciple

THE DISCIPLE IS ANYONE who is sufficiently motivated by desire for direct religious experience to put up with the demands of the master. Unlike the leaders of contemporary cults, who relish attracting followers, spiritual masters often place extraordinary obstacles in the way of disciples. One of the most famous initiatory narratives in all of religious literature, which may or may not be factual, is the initiation of Milarepa by the Tibetan Buddhist master Marpa. The initiation was severely complicated by Milarepa's extensive and sinful background in black magic.[19]

Milarepa, on hearing the name Marpa the Translator, became "filled with an inexpressible feeling of delight" and went looking for the lama, hoping to receive instruction. He saw a tall monk plowing a field and identified himself as a sinner who was looking for Marpa. The monk told him to continue the plowing as he went looking for the teacher. That was the first in a series of tests that nearly broke Milarepa's body and soul. The trials included the following:

- Make a choice between instruction and material support
- Perform a number of magical tasks
- Undo the magical damage previously done

- Build a tower, tear it down, return all the stones to their place
- Repeat this four more times, with different designs for each tower
- Repeatedly offer gifts to Marpa, who will repeatedly refuse to accept them and will toss you out of the house
- Build a shrine and a covered walkway leading to it
- Put up with a severely injured back
- Accept the master's contemptuous disregard

Milarepa did obtain extensive help from Marpa's wife, including assistance with hatching a plot to receive instruction from another lama. But Marpa saw through all these designs and undermined them at each turn. We do learn, however, that when Milarepa gave up and went away, the master sat still, shedding tears and praying for his return. It was only after Milarepa prepared to leave for the final time, without having once shown or felt any anger toward the master, that Marpa finally gave him initiation. Explaining that nine great trials were necessary to erase the sins of black magic, Marpa added that he had also been testing Milarepa's zeal, perseverance, wisdom, and compassion—as well as whether he could completely disregard the comforts of this life. Only such a demonstration made Milarepa ready for instruction.

Considering that the entire relationship between master and disciple is a transformative and difficult quest rather than membership in a club, it is sensible for the master to launch his disciple's program with a difficult initiation. The original test is thus essential for exposing the necessary virtues for the disciple's long-term success. These include confidence and humility, benevolence, compassion, ability to forgive, self-control, and steadfastness. In fact, the more difficult the test, the more engaged the master himself must be, thus revealing his own compassion and grace as he lays open the student's potential strengths. Through initiation, the ideal disciple, like Milarepa, exhibits those qualities that are most highly prized in the mystical tradition to which he belongs. In India, for instance, those qualities are reflected in the words of the Bhagavad Gita, which stipulates that the disciple is to perform actions that are "free from all attachment, performed without passion and without hate."

One of the best-known Indian narratives on a guru and disciple illustrates this kind of unselfish service taken to an extreme degree. It is a Mahabharata episode involving Karna and his teacher Parashurama. One day when the guru was asleep with his head in Karna's lap, an insect began to gnaw on the student's thigh, causing extreme pain and bleeding. Karna remained perfectly still, until his dripping blood finally woke up the teacher.

The initiatory process in numerous cultures around the world is familiar to anthropologists. Victor Turner has discussed initiatory pain, humiliation, and social manipulations in great detail. The core of his argument is that initiation separates the novice from his previous social group and creates a temporary identity—one characterized by an inverted or even effaced sense of self. Where a whole group of initiates are involved in the process, a unique and temporary society emerges—"communitas"—a sort of brotherhood or sisterhood.[20]

Many of the oddest features of spiritual initiations by spiritual masters can be understood along these lines. For example, what may seem like no more than service rendered to the master can take astonishingly radical turns in the initiatory phase of the relationship. In his study of the master-disciple relationship, Abdellah Hammoudi recounts the story of Sidi al Haj 'Ali, founder of the Moroccan zawiya of Ilgh, who, as a young man, abandoned the comfortable life of a family man, along with his studies, to become initiated by the Sufi master Sa'id ben Hammou.[21] This entailed, at the very least, five years of wandering from town to town, wearing tattered clothes and displaying a large-bead rosary and a fakir's cane. It also meant enduring the humiliation of begging for food in the market, fasting, putting up with his family's judgment that he had lost his mind, actually losing his mind (depersonalization), renouncing virtually all social relations, suffering from sleep deprivation, discomfort, and pain, and completely obeying his master's every demand. None of these trials, according to 'Ali, would have amounted to much without prayer, for which the disciple desperately needed his master. Then, too, the master had to be approached with utter submission.

But austerity in one form or another appears to be a defining characteristic of both the spiritual master's spiritual quest and the initiation into a circle. It may be poverty, fasting, sleeplessness, or even self-inflicted pain. For example, the Islamicist Annemarie Schimmel has noted that the disciple must demonstrate his ability to follow in the path of his master by displaying "pain (*dard*) through love (*'ishq*) as the most efficacious and rapid means of attaining perfection (*kamal*) and/or union (*wisal*)."[22] This self-sacrificing preparation extends beyond Islam to every tradition where spiritual masters attract initiates. And as seen in the previous chapter, the austerities not only are moral displays but also act as psychological and even neurological techniques in the mystical practice.[23]

The profoundly asymmetrical relationship between the master and the disciple, the total loss of the disciple's autonomy and dignity—all in the pursuit of love (joy) and enlightenment—calls for dramatic tropes. This is true in virtually all the traditions and languages where such relations prevail. For example, in a society where gender determines power (males have it, females do not), the disciple is feminine. But paradoxically, in such cultures the master is also in some sense maternal. So the master either injects knowledge into the disciple, like male seed, or nurses him like a mother. In Sufi discourse, for example, just as the infant drinks milk at the breast of its mother or wet-nurse, receiving from her the sustenance without which it would perish, just so does the infant of the spirit drink the milk of the Path and the Truth from the nipple of prophethood, or the wet-nurse of sainthood (*wilayat*), receiving from the prophet or the shaykh—who stands in place of the prophet—the sustenance without which he would perish.[24] In contrast, the transmission of prayer (*dhikr*), which is so central to the progress of the disciple, is likened to the injection of seed.

Similarly erotic images pervade Hasidic writings taken from Kabbalistic mystical literature and transformed into a model of their own social bonds. The following are two examples of this pervasive approach:

> The pleasure that a whole-hearted man finds in his study and prayer, who attaches himself to the sacred letters, to holy souls (Job 19:26 [which states that]) because God has seized hold of my flesh like an aroused member during sexual intercourse, which is the highest pleasure, just so a wholehearted man who enjoys his study is considered alive. Whereas if he does not have such pleasure he is like an unaroused lifeless member, although such a man might have fear. It is necessary that he feel both love and pleasure.[25]

The Besht, too, is quoted as stating, in Keter Shem Tov (p. 4a, no. 16) in commentary on that verse (Job 19:26: "From my flesh I shall see God"): "Just as you cannot sire [a child] in physical copulation unless your organ is 'alive' and [you are filled with] desire and joy, so it is with spiritual coupling, that is, with regard to the words of Torah and prayer: when it is done with a live organ, in joy and pleasure, then you can be fecund."[26]

Bharati Ma, the Indian teacher and holy woman, calls her followers and visitors son or daughter and they call her mother (*mataji*). She explains: "I will never have disciples, only 'children,' because that is the way a real guru should treat a disciple: as a spiritual son or daughter. And the bond between them is far more intense than that between a physical parent and a child."[27]

Clearly these are just metaphors for the joy and love that underscore successful discipleship in the spiritual quest. In chapter 5, I discussed the transition from organic societies to moral communities—early Christian and Buddhist ones, for example. I also discuss the martyr Perpetua in the last chapter as someone who abandoned family relations for a new type of group, one based on faith and friendship.[28] The master and his group of disciples constitute an extreme example of a moral community, but it is a group that feeds on the metaphors of organic (familial and marital) love in order to emphasize its new dispensation. As a result, we should not take these images as a hint of tantric-style sensuality. But we should also recognize that the use of erotic tropes for the spiritual process is not arbitrary either; it does lead to, or is somehow connected with, potentially antinomian relationships based in a carefully calculated disregard for respectable norms and conformity to "safe" modes of social behavior.

Love Against the Law

The passion and spontaneity of the spiritual initiation—often manifest in the behavior of the master himself—can spill over into odd behavior. Religion scholars call such behavior, whether it is the use of striking literary metaphors or actual conduct, "antinomian." It violates, or appears to violate, the norms that guide proper conduct in the religious society at large. For example, R. Nahman of Bratslav (*Likkutei Etzot*, Zaddik, no. 57) writes about the tzadik who evaluates potential disciples: "At other times he conceals himself and hides from them so that they are unable to come near him. He may also provoke questions about his objectionable and astonishing behavior, to the point that their mind becomes twisted and confused because of his strange conduct with them."[29]

In a similar vein, while it is known that Islamic law (Shari'a) frowns upon certain types of entertainment—especially if women are going to become involved—Sufis have sometimes thumbed their nose at such injunctions. Sultan Valad, who was Rumi's son, describes the rituals of *sama'* practiced in Sufi lodges, consisting of both lively music and dancing (no women). Still, only those who have undergone years of spiritual practice in self-restraint can participate in borderline rituals. There have been Sufi masters and disciples who have gone even further in testing legal norms of purity and proper conduct. The so-called *muwallah* (mostly in Sunni Syria) represented a hybrid of mystic, saint, vagabond, and healer, a man dressed in filthy clothes who appeared to avoid praying like other good Muslims. Such men included Yusuf al-Qamini (d. 1259), Shaykh 'Abd Allah al-Fatula (d. 1301), Jandal b. Muhammad (d. 1277), and perhaps even the great Andalusian mystic Ibn al-'Arabi (d. 1240).[30]

In India, holy men have at times engaged in odd or eccentric behavior, although this is usually cited to illustrate not their disregard for the law but the fact that they have moved beyond all legal and moral distinctions, such as right and wrong or pure and impure:

> Many are the ways of Siddhas. I knew a great Siddha, Zipruanna, who was always naked. He would lie on a heap of garbage. He ate whatever passersby gave him. Although he sat surrounded by filth, he was never affected by it. He saw only equality everywhere. He was always happy. Although he was in the body he knew that he was totally different from it. He was an ecstatic Siddha.[31]

There is something serious about the odd and seemingly irreverent behavior of spiritual masters. At the very least, their behavior exposes the tension between religious law and the institutions that uphold it, on the one hand, and devotional love, on the other. It challenges, comically or seriously, religious and legal norms as a source of religious authority and as a way of binding people together into a group. Of course, this can be dangerous business. Although these groups are not actively threatening and most Sufi, Hasidic, Hindu, and Christian ecstatic groups tend to be pacifistic, they often draw acts of violence in their own direction. A perfect Christian example of this would be the early Quakers led by George Fox, James Nayler, and a few others. These charismatic groups followed only the indwelling Light of Christ and intentionally disregarded religious sacraments, laws, and ministries. They willfully brought charges of blasphemy against themselves and went to prison, putting up with extreme physical hardships. John

Audland put it in the following terms: "Being faithful to the Light will lead you to the Death upon the Cross."[32] In other words, their persecution was a sign of authenticity and did not call for defense or counteraggression.

This willingness to absorb violence could go hand in hand with an irreverent attitude toward religious law by instigating, not lewd behavior, but rather a sort of holier-than-thou ethos that went far beyond the norm. This hyper-morality could have the appearance of antinomianism in challenging existing standards. For example, this may be the reason the Ḥasidim in Europe, until the Rebe's court began to institutionalize their conduct, valued constant movement and poverty. The nonconformity of the Hasidim points back to the influential but antinomian mystical Zoharic text, Re'aya Meheimana ("Faithful Shepard"), which appears to have absorbed the Franciscan emphasis on poverty as a mystical tool, an odd notion for mainstream Judaism.[33]

It is not easy to say why this contrarian attitude would emerge in the context of mystical and devotional groups. Some mystics within monastic cloisters remain faithful to the rules, and some nonmystical groups stretch the rules in other ways. Victor Turner, as we have seen, argues that all initiations create temporary groups of initiates who exist in a universe with separate rules. These "liminal" groups often invert the values that govern ordinary existence, and we may be seeing this social dynamic at work with the master-disciple groups as well. But there is a marked difference: while rites of passage and initiation are temporary and ultimately reinforce enshrined norms, the spiritual group with its charismatic master can be a direct threat to religious authority. The group that surrounds him can make a huge political difference.

There are several reasons for this: the spiritual master who becomes engaged in the world possesses an enormous capacity to mobilize his followers. His authority is beyond question, while his moral convictions, for which he relies on direct revelation, present a clarity and simplicity often lacking in ordinary laws. His messianic or utopian goals can be compelling. Furthermore, the shaykh/tzadik/guru does not acquire his magnetism from holding some office within an institution, or from code or statute, but from his spiritual insight and his love—his charisma. These qualities bypass hierarchies and established institutions; they are politically explosive. This is how Max Weber put it: "Pure charisma does not known [*sic*] any 'legitimacy' other than that flowing from personal strength, that is, one that is constantly being proved."[34] Finally, the austerities and deprivations associated with membership in the master-disciple group are excellent preparation for the self-sacrifice—even martyrdom—that may accompany the political struggle.

All of this can go awfully wrong. The classical example for the political dangers of the devotional group in Judaism, and one impetus for the rise of the small Hasidic group, was the catastrophe of the false messiah Sabbetai Tzvi (1626–76). Tzvi was a Turkish-born Kabbalistic mystic and ecstatic who began his career by attracting a small number of disciples. He was declared the messiah by Nathan of Gaza (1643–80), an influential Palestinian Jewish theologian. Eventually Tzvi drew a huge number of followers in his travels throughout the Near East and Europe, Jews who became convinced that the messianic era had dawned. Their intention was to return to the promised Holy Land. The entire episode ended with Sabbetai Tzvi converting to Islam under a threat—his other choice was death. His decision to avoid martyrdom marked him not only as a false messiah but also as a rather shoddy example of the spiritual master.[35]

Another Tzvi, but the very opposite of Sabbetai Tzvi (if one may speak about spiritual leaders in terms of opposites), and a man who also focused his love on the messianic Zion, was Rabbi Tzvi Yehudah ha-Kohen Kook, who connected mystical love (for God, but especially for the land of Israel) to political activism.

Rabbi Kook was the son of the eminent scholar and first chief (Ashkenazi) rabbi of Jewish Palestine, Avraham Itzhak Kook. He continued his father's work at the Merkaz Harav Yeshivah, but after the 1967 war he also became the most prestigious spiritual proponent for the reclamation of the entire land of Israel and settlement on the West Bank. What made Rabbi Tzvi Yehudah Kook different from simple nationalistic rabbis was the profoundly charismatic relationship he had with his student-disciples. He was a great lover of his students, whom he often treated as though they were his children, and the love he received in return also transcended the ordinary rabbi-student relationship. In addition to his formidable knowledge, Rabbi Kook was credited with the power of prophecy, usually linked with remarks that seemed to predict the painful 1973 war. Furthermore, Rabbi Kook was an ascetic who deprived himself of food and sleep. Although his students denied that he was a mystic—he was not a Hasidic tzadik—many of his later students did treat him with that level of deference, and many others claimed that he had the power to perceive the hidden nature of reality.[36]

The political effects of small devotional groups can far exceed their actual size and number. To give one example, in South Asia, Islam spread largely with the help of Sufi saints (*pirs*) who led military groups into some of the most remote areas. Many of these fighting Sufis became martyrs (*shahids*). Their tombs (*dargahs*) can still be seen all over the subcontinent, where they continue to attract devout followers. Some of these sites became influential regional centers for Islamic culture and political activism. The Sufi bands finally disappeared under British colonial rule.[37] However, it was the rural devotional Sufi group that later provided the strongest impetus for resistance to the British (and the rise of Pakistan), despite the fact that such a vast and modern national struggle required a far larger—indeed, national—organization. It was often the Sufi shaykh—for example, Jama'at Ali—who mediated between the actively militant local devotional groups and the nationwide organization that gained traction for their efforts. For example, in 1946 Jama'at Ali gave politics a religious legitimacy by declaring that Jinnah (Pakistan's leading statesman) was "an intimate of God (walli Allah)."[38] He had decreed earlier, in a revivalist state of mind, that any Sufi follower of his who did not read the magazines *Al-faqih* and *Risala*, which promoted the proper religious, economic, and political agendas, would be "barred from any hope of intercession in this world or on Judgment Day" through the shaykh.[39]

It is important to note that scholars of Islam tend to see the spiritual struggle of the Sufi (master-disciple) group as complementary to mainstream legalistic Islam. Arthur Buehler, for example, argues that they are mutually reinforcing: Sufism invigorates the institutional religion by giving life and vigor to its major symbols. The inner jihad is a deepening of the outer jihad.[40] Even when a Sufi group is actively undermining the existing government of a Muslim country, it is invariably in the name of a purer interpretation of Islam.[41] In other words, Sufi groups are never simply nihilistic.

Furthermore, the Sufi social order can act as the model for a much broader—even national—society. According to Abdellah Hammoudi, this is precisely the case in Morocco. There the shaykh-disciple relationship has become the model for all power relationships.[42] This is particularly true in reference to the king—a direct descendant of Muhammad. The king is the symbolic shaykh in relation to the entire nation's figurative discipleship, and every other relation of power in Morocco participates in these twin poles of control and submission. Annual rituals, festivals (*'achoura*), and liturgy perpetuate these authoritarian values and play them out in actual practice. This does not imply that Morocco is a dangerous country, of course, but it does demonstrate the enormous appeal of a religious idea and social arrangement based on love and devotion. And even if it is unlikely that such a model would be extended to other Arab countries (including Jordan, which is also run by a fairly well liked king), this still demonstrates the possibilities inherent in a highly persuasive and satisfying social and religious arrangement.

Religious Groups at War

WESTERN SECURITY GOALS PRESENT Islamic terror as a grave threat to the world's security and prosperity. It has become a bit too easy to put all the various groups that engage in terror and military activities in one bag and ignore important distinctions between them. With regard to a future nuclear threat, the main distinction that must be observed is between the aggressive and victory-minded, on the one hand, and the self-annihilative struggle, on the other. The first seeks concrete worldly gain, while the second agrees to undermine all pragmatic goals for the sake of a spiritual utopia. An example of the first is the Muslim Brotherhood in Egypt (Ikhwan al-Muslimun), which has been participating in elections in order to gain political power and ultimately promote Islamic rule in Egypt.[43] Naturally, this group seeks to fight Western influence and change Egypt's policies toward the West and Israel, and it accepts violent means of doing so. Similar observations can be made about Hizbollah in Lebanon and Hamas in Gaza and the Palestinian territories. These organizations are extremely aggressive and dangerous, and even though they recruit, train, and deploy suicide terrorists, they are not self-annihilative. There are similar groups around the Muslim world that seek to establish Islamic government and promote the interests of Muslims.

In contrast, I believe that small-scale groups that center on a spiritual master—a Sufi shaykh, for instance—are potentially annihilative. Most Westerners who know about Sufism through the poetry of Rumi or the dances of the whirling dervishes or Pakistani music are surprised to learn that Sufis can be militant. In fact, Sufis have a very long history of not only political engagement but also warfare. Their goals in taking up arms have ranged from fighting Christian crusaders to spreading Islam in the wilderness of South Asia and resisting corrupt governments in Ottoman Turkey.[44] Today there are numerous Sufi groups that have taken up arms as insurgents in Iraq and elsewhere in the Muslim world.[45] The nature of the master-disciple group, which is built on charisma or love, guarantees that the struggle of the Sufis is not directed at creating institutions but rather at radically unmaking what they perceive as corrupt and incorrigible. In other words, their goals are annihilative, and they are not concerned about their own survival.

The same distinction—between aggressive groups with concrete material goals and self-annihilative groups—can be observed in the Jewish and Hindu cultural landscape, but with a difference. Where a militant Sufi group might seek to undermine the political center of its country and is willing to self-destruct in the process, the Hindu and Jewish groups are trying to strengthen allies in the political universe. They are "right-wing" activists within a democracy. The danger they pose emerges under extreme conditions when their goals are crushed, such as during Israeli withdrawal from Palestinian territories or India's potential retreat in the face of Muslim demands (over Kashmir or gaining minority rights in India). In other words, the nontriumphal Islamic devotional-political groups are *inherently* destabilizing, while the Jewish and Hindu groups become destabilizing when they fail to rationalize defeat and turn to internal violence for those reasons only.

For example, in Israel the most notable examples of charismatic groups are Gush Emunim (founded by Rabbi Moshe Levinger and inspired by Rabbi Tzvi Yehudah Kook) and a number of splinters that have emerged from it, such as Hilltop Youths (led by Avi Ran and inspired by Rabbi Shaul Halfon) and even the quietistic settler groups led by Rabbi Avi Gissar. The man who assassinated Israeli prime minister Itzhak Rabin, Yig'al Amir, also emerged out of the nationalistic-rabbinic milieu, where he may have been directly inspired by Rabbi Nahum Rabinovitch or Shmuel Dvir. A short time before the assassination on November 4, 1995, a man named Yosef Dayan allegedly performed a ritual known as Pulsa Di'Nura—which is a pseudo-Kabbalistic curse. The ritual was directed at Itzhak Rabin, who was accused of treason for signing the Oslo Peace Accords. Another man, Avigdor Eskin, has made the same claim. Neither of these men, especially Dayan, recognizes the legitimacy of the Israeli political order, and both seek to establish the ancient Davidic monarchial order (Malkhut Israel). A small example of the mischief such groups are capable of instigating was provided during the withdrawal from Gush Katif (Gaza Strip) in 2005. The violence that erupted there demonstrated how aggressive Jewish settler groups can turn into immolative ones if they are pressed to abandon a settlement.[46]

The situation in India is similar to some degree. Religious organizations and political parties have staked their claim on the national agenda. Some of these groups are led by individual gurus with varying spheres of influence, but the sensible ones—for instance, Gayatri Parivar and Bharata Mata Temple—have maximized their interests by banding together under such umbrella organizations as the Vishva Hindu Parishad (VHP).[47] The agenda of Hindu nationalists has taken aim at issues such as minority rights, Kashmir, promoting the economic interests of the upper classes and castes, and the conflict with Islam over contested places of worship. Furthermore, the prestige of the guru and his immense power over his followers—or even an entire region, such as the Kanchi Shankaracharya in southern India—guarantee the increasing role of the guru in Indian politics. However, like nationalistic rabbis in Israel, guru-centered, nationalist Hindu groups are overwhelmingly pragmatic. In some ways, they resemble right-wing evangelical groups in the United States—such as those led by Jerry Falwell or Pat Robertson—during a sympathetic presidency.

As I indicated earlier, the danger associated with the Jewish and Hindu religious militants is the threat of the failure of their nationalistic agenda, which is religiously conceived. Certainly, like the Jewish groups in Israel, the Hindu groups may on rare occasions engage in aggressive violence against external enemies. For instance, VHP members were involved in the Provincial Armed Constabulary, which attacked Muslims in Meerut in 1987 (*India Today*, June 15, 1987). However, as Lilly Weisbrod showed with reference to militant Jewish groups, the real danger of a master-disciple group that grows within a religious majority of a functioning democracy, whether Jewish or Hindu, lies in how it copes with failure when the national agenda turns to compromise rather than total victory. That is why both Itzhak Rabin and Mohandas K. Gandhi died at the hand of religious-nationalistic terrorists, who emerged from small groups within the religious majority.

Two central points must be emphasized relative to all of these groups. First, we must distinguish between apocalyptic self-annihilative groups and other such groups that attempt to establish religious laws and build institutions that will contribute to public welfare and promote the group's self-interest. When the latter groups are aggressive (which is not always the case), they are triumphalist rather than intentionally self-destructive. I believe that such groups can ultimately be coaxed back from the brink of wholesale annihilation because they are invested in pragmatic goals. This is an entirely distinct goal from the goals of groups that do not believe that such world-building is possible or desirable and that seek merely to undermine what is already in place for the sake of a messianic or apocalyptic utopia or heavenly existence. These groups may actually try to provoke a nuclear or other massive confrontation. However, to repeat a critical point: when the goals of even the more pragmatic group are frustrated, it may at times revert to the tactics of the messianic group.

The second point to keep in mind is that most of these groups are engaged in a battle, first and foremost, against the host culture in which they operate. Although the "great" enemy may ultimately be the foreigner, the immediate target in sight, so to speak, is the status quo and its supporters. This target may include the secular ruler who negotiates with the enemy (for example, the presidents in the Arab or Pakistani states who are backed by American aid, or the Israeli or Indian governments), the state as a whole, or even the mullahs who rule an Islamic republic (and the rabbis who serve in the Israeli government).[48]

Finally, I do not believe that the Indian and Israeli groups will be able to break into the secular political-military establishments that control the significant weaponry. What they could achieve in terms of aggressive or self-immolative behavior is an enormous provocation, such as the bombing of the Al Aqsa Mosque in Jerusalem or a roughly equivalent symbol in South Asia.

Conclusion: The Warning Signs

In the event of a nuclear confrontation between Israel and Iran or between India and Pakistan, or even in a lower-level crisis involving thousands of conventional rockets (Israel, Lebanon, Syria, Gaza), experts will certainly examine the enemy's ability and willingness to withstand great pain. They will look at economic wealth, military strength, physical infrastructure, political stability, international support, and many other factors. Where should religious considerations fit into such assessments of the enemy? How do experts decide whether religion is more likely to make another country launch a potentially suicidal attack?

I believe that the critically important answer to these questions may surprise some observers. There are currently numerous theories that predict which religious dimension may lead to an all-out conflict in regions such as the Middle East. Most of these theories—proposed by Mark Juergensmeyer, Manus Midlarsky, Thomas Robbins, James Waller, Ross Moret, and others—focus on apocalyptic beliefs and on the demonizing of the other side in the light of a final cosmic battle. Other theories, such as Bruce Lawrence's, play up the moral absolutism of a fundamentalist approach to religion.[49] While such theories are valid in their assessment of the intellectual and ideological risks, I believe that the more dangerous factors are emotional and organizational. Furthermore, I do not believe that the greater threat will emerge from a theocracy (even a "fundamentalist" one such as the mullahs of Iran) or from a religiously authoritarian organization (Hamas) or state. On the contrary, I fear a state where the center is weak. The internal threat to such a state, if it comes from a master-disciple group, may act as the trigger to a nuclear attack, or the attack itself may somehow be carried out or inspired by members of this charismatic group. In this chapter, I have listed and discussed the principles that may operate in such a case. If my hypothesis is correct, then there are several warning signs to look for in advance of a crisis. The first area to examine is the threat to the state's institutional center, whether it is secular (Israel, Syria, Pakistan) or legal-religious (Iran). How many devotional groups are there in the country, and how much influence do they have? Assuming that potentially destabilizing religious factors can be identified, the following major signs should be taken into account in relation to each group:

1. How mystical or ecstatic is the leader?
2. How prominent is love discourse within the group?
3. How reformist is the group's outlook? How strong is the moral discourse?
4. How new is the group?

Each of these questions can be subdivided into subsidiary issues and signs. For example, under the second heading, we would also try to identify signs of initiatory process—the more rigorous the initiation, the greater the potential danger. Under the third heading, we should look for evidence of conflicts with the prevailing authorities (trials, jail time, and so forth). What we are trying to discover is how charismatic the leader of the group may be. If the group is older and has established institutions and a bureaucracy it is less dangerous than a group that hinges on the loving relationship with a leader. A group that is both ecstatic and reformist is more likely to threaten the foundation of the host state. It is important to keep in mind that any action such groups may instigate will be directed against internal enemies, not a neighboring country. An attack on the neighbor may simply serve this broader campaign. As a result, it may prove unrealistic to attempt to deter the group by threatening the whole state with destruction or loss of life.

The question of metaphysical belief may also come up: Is the group messianic, apocalyptic, or prophetic? How much does it value worldly prosperity relative to metaphysical joys? But this question is not centrally important. The calculus of cost-benefit may be influenced by views on the afterlife, as we shall see in the next chapter, but this factor pales compared with the overwhelming rewards of satisfying a beloved leader or the dynamics of social emotions based on the separation between inside and outside (us/them) within the broader national setting. The desire for radical social-religious reform and the utopian visions would amount to very little without the pleasures of charisma and love.

Finally, and to repeat, I do not believe that the mullahs of Iran represent the mortal danger often attributed to them. Nor would any religious party in Pakistan that might take over the state, or a radical devotional group in India. The Hizbollah in Lebanon and Hamas and Islamic Jihad in the Palestinian territories fall into the same category. These are all dangerous players, bent on destroying their enemies. But they are also interested in governing, in maintaining institutions, and in furthering their own power and influence. They are not annihilative (ecstatic) but subject to deterrence or some other form of rational negotiation. The greater danger lurks among the far smaller groups that would seek to undermine these larger religious (or secular) parties that represent the status quo.

Most Americans would find it odd that I consider Hizbollah and Hamas representatives of the status quo. But both organizations have developed elaborate bureaucracies, public works, and educational and economic institutions, and both have a growing interest in seeing these social constructs prosper. But somewhere in Lebanon or elsewhere in the Middle East, very likely in Iraq or Afghanistan, there may be a young Sufi mystic who will become a master (shaykh) with a small but growing number of fervent disciples. This mystic might decide that Hizbollah or one of the other known organizations has sold out for worldly gain. One of his secret but most ardent followers may be a high-ranking military official with access to a weapon of mass destruction and no stake in the world around him. He will consider himself a martyr for a good cause.

Such a violent splinter group operating in the Middle East or elsewhere in the world cannot be fully understood without looking at one additional factor: faith in heaven. It is not that getting to heaven seems irresistible and drives people to suicide. Instead, as we will see in the next chapter, faith in heaven correlates with social arrangements and makes the believer miserable. This can be an incendiary mix.

CHAPTER 8

The Paradise Paradox: The Misery of Heaven-Addicts

Every one that hath forsaken houses, or brethren, or sisters, or father, or mother, or wife, or children, or lands for my name's sake shall receive a hundredfold and shall inherit everlasting life. . . . Enter upon the road to the Holy Sepulcher; wrest that land from the wicked race, and subject it to yourselves. That land which as the Scripture says "floweth with milk and honey," was given by God into the possession of the children of Israel . . . the land is fruitful above others, like another paradise of delights. . . . Accordingly undertake this journey for the remission of your sins, with the assurance of the imperishable glory of the kingdom of heaven.

—Pope Urban II, speech at Clermont, 1095, announcing the first crusade[1]

He (Allah's Blessings and Salutations may be on him) also said: "The best of the martyrs are those who do *not* turn their faces away from the battle till they are killed. They are in the high level of Jannah [paradise]. Their Lord laughs to them (in

> pleasure) and when your Lord laughs to a slave of His, He will not hold him to an account.
>
> —OSAMA BIN LADEN,
> "DECLARATION OF WAR AGAINST THE AMERICANS," 1996[2]

HEAVEN AND VIOLENCE HAVE ALWAYS been a close fit; the quotes opening this chapter are but two examples. How much more likely is someone to start a war or to turn down a political compromise if his battle leads either to victory or to heaven? Belief in heaven, in other words, has made a lot of innocent people unhappy. But what is less obvious, and far more insidious, is that faith in a wonderful afterlife can make the believer himself less happy as he awaits the end. I am aware that this is counterintuitive. After all, people invented heaven—or believe in heaven—to make a short and sorrowful life more bearable. Heaven, seemingly, is humanity's hope in the face of the abyss. So I must ask readers to bear with me until the very end of this chapter as I explain why a joyful afterlife is counterproductive in terms of its effects on one's well-being in this life.

According to a 2004 poll by the Pew Forum on Religion and Public Life, belief in heaven as a real place to which one goes after death is extremely pervasive in the United States. Seventy-four percent of all Americans believe in an afterlife, with evangelicals coming in at 86 percent; Mormons are highest at 95 percent, and Protestants in general polled at 84 percent. Among the major religions, Buddhists score lowest in faith in heaven at 36 percent, and Jews are next at 38 percent. A very high percentage of Muslims in America believe in heaven (85 percent), and Hindus in America less so (51 percent).[3]

It is not clear how the Pew researchers depicted heaven for the sake of measuring attitude, but for the sake of the widest possible recognition, it might have been something like the way the Christian Broadcasting Network (CBN) put it: "What is heaven like? It is better than anything any human being could ever imagine when he tries to picture the best thing that could ever happen to him. Beyond that, there is not much else that we could say!" However, on his Web page (CBN.com), Pat Robertson still manages to say a good deal about heaven: "We read in the Book of Revelation of streets of gold that are like clear crystal. It needs no light because the light of God is in it. It has a river flowing from the throne of God, with trees growing on either side of it that are for the healing of the nations. . . . The Bible says, 'There shall no longer be any death; and there shall no longer be any mourning, or crying, or pain.'"

We have seen in chapter 4 that religious techniques for promoting happiness begin by offering a critique of simple replenishment pleasure—the treadmill pleasure, as it is often known. As a result, a "permanent" value replaces the sensual object of desire; as Rick Warren describes it, the purpose-driven life is the fully satisfying one. It is purpose, on this understanding, that cures the addiction to constant novelty and an ever-intensifying need for satisfaction. However, it is not clear that the concept of purpose solves the problem when "purpose" means simply projecting onto the future something similar to what we wish for now. In other words, a purpose that does not imply a radical change in the way that individuals seek the fulfillment of desire is just another expectation.

The idea of heaven belongs here—that is, in this religiously sanctioned habit of deferring present desires onto a distant future. Heaven does in macrocosmic time what "purpose," "goal," "summer vacation," "promotion," "it will make sense later," or "let's see what's on TV after dinner" do for microcosmic time. It bolsters a mental habit of expectation that perpetuates the problem that seems to ail the simple present moment. Oddly enough, this is not an inevitable way of being religious. In fact, the best critique of this hedonic deferment and an incisive theory of what makes it so compelling comes from those religious/philosophical traditions—Hinduism and Buddhism—that reject both the concept of expectation and its final target: a permanent and wonderful heaven.

Indian Critiques of Deferment

ANCIENT INDIAN PSYCHOLOGISTS REJECT the concept of the wonderful and lasting reward—heaven—and its essential foundation, the individual self. Sure, they too have criticized the hedonic treadmill, but their understanding of pleasure is far broader than the sensory or addictive varieties of pleasure. To the Hindu and Buddhist thinkers, pleasure covers anything that is evaluated positively. Like the positive side of a magnetic field, it is merely "this" side in a dualistic habit of mind, with pain on the other side. The Bhagavad Gita (2.38), to give one example, says that one should move beyond *all* oppositions such as pleasure-pain, cold-heat, good-bad. To put it more abstractly, one should renounce the mental habit of splitting everything in the world into opposites. Why is this? As Indian thinkers conceive it, opposition is not a true feature of the world but the way in which human awareness imposes structure, with the misguided influence of emotions such as fear and desire on the straying mind. We could almost say, anachronistically, that mental oppositions result from the biology of adaptation and the evolution of the brain.

Other Indian writers over the centuries have explored this hedonic (positive/negative) theme in great detail. For example, the psychological text Yoga Sutras of Patanjali from the second century CE explains how the patterns that define the personal self (memory, anticipation, core beliefs, deep feelings) congeal around our ingrained habits of clinging to some things and feeling repelled by others. This idea remains very influential among Indian thinkers, even popular ones. Most recently, the Indian mystic J. Krishnamurti has linked this same habit of the active self—the self that we so dearly believe ourselves to be—to the fear of our own death. Discussing the fear of death, Krishnamurti writes: "So the thing that we are afraid of losing when death takes place is the structure that thought has built as 'me,' the form, the name, and the attachment to the form and to that name, which are pain, pleasure, anxiety. All that is the 'me,' the 'you.'"[4]

It is fascinating to see in Krishnamurti's incisive examination just how vital the future is in the way the "me" constructs itself. But it is not necessarily just the distant future of our final demise that concerns the self. It is the passing away of any given moment in time—it is time itself. "So you have to find out whether time has a stop. Time has a stop when there is no longer the movement of that stream. That stream is fear, that stream is conflict, that stream is sorrow."[5] Certainly we are sad at the death of our loved ones and afraid of our own. But we are more deeply anxious about losing permanence itself, an anxiety that permeates every single instant in the flow of our time. We want not only love, family, house, and career but also the idea of these things to hold still for a while. Passing time feels like death to the self—millions of momentary dissolutions of whatever we are holding at one moment, then lose in the next.

What appears to hold it all together, which Krishnamurti is trying to undo while following in the footsteps of older masters like the Buddha himself, is the clinging self, with its anticipation, memory, beliefs, and feelings.[6] By renouncing opposites, one can break the habits of mind they instill and thereby step outside the stream of time. Or at least one learns to let go of the self within the stream. This brings total freedom—the foundation and goal of Hindu and Buddhist psychologies.

To summarize what Indian thinkers have to say about a future heaven:

- Heaven perpetuates a false sense of the self.
- Heaven destroys our capacity to be free and individual.
- Heaven kills the truth and joy of the present moment.
- Heaven, in sum, makes us less happy.

Some contemporary Christian theologians would gladly agree. Paul Tillich, to give one example, prefers to think of heaven in microcosmic (or altogether nontemporal) terms: "Eternal Life," which refers to what is normally known as the individual afterlife, "means that the joy of today has a dimension which gives it transtemporal meaning."[7] This sentiment could almost (but not quite) have come straight out of India's ancient philosophical traditions, as we see by comparing it with Krishnamurti. But Tillich is hardly representative of what most people today think about heaven—even American Hindus.

In fact, there is probably too much working against the Indian critique of heaven. There is, first and foremost, the biological drive, feeding on dopamine-enabled cravings that keep us moving forward. There is also the psychological and even cultural prizing of the individual self that has a future. It is all there in the future; you can replace the term "purpose-driven life" with "future-driven life" and see what I mean.

We could argue that belief in heaven is silly or that it is dangerous because it justifies violence and makes us passive in the face of injustice.[8] And it is easy to see how such arguments would proceed and what the counterarguments would look like. But I believe there are far more interesting things to say about heaven. Following the lead of the Indian psychologists, we can argue that faith in heaven is an actual force that shapes our view of the personal self and makes that self future-oriented rather than grounded in the present. To put it more broadly, heaven is part of a broader intellectual and religious habit of mind that makes our life less fulfilling and ultimately less real. Heaven depletes life of its potential joy. I call this the "paradise paradox" (with apologies for the alliteration), and it is the topic of this chapter.

The Afterlife

THE AFTERLIFE IS A creative writer's fondest dream. She can make it anything at all, paint it with the most vivid images of joy or torment, and who will call her a liar? A group of four special-effects artists won the Oscar in 1999 for best visual effects in a movie called *What Dreams May Come*. In this film, heaven is better than the grooviest, hippest psychedelic LSD trip, with a satisfying psychological twist.

Given how creative one can be with the afterlife, what is most surprising about three millennia of world religious literature is not the diversity of heavens and hells but the fact that predictable patterns in conceptions of heaven and hell do in fact emerge. The first and most obvious is that death, against all evidence to the contrary, is not the end. The second is that the afterlife has something to do with righted wrongs. And finally, heavenly rewards feel good—heaven is hedonic. Our own Western views of heaven originate with biblical notions, with a few roots in Persian

(Zoroastrian) and Greek ideas.[9] The Hebrew Bible is already eclectic about heaven. (Hell will be excluded from this discussion.) The views that remained most influential when Christianity emerged were apocalyptic-national (the Books of Ezekiel and Daniel) and individualistic-poetic (Psalms 49, 73). The first vision of the afterlife was about bodily resurrection and a beatific existence in a divine kingdom on earth: "And they shall dwell in the land that I have given unto Jacob My servant, wherein your fathers dwelt; and they shall dwell therein, they, and their children, and their children's children, for ever; and David My servant shall be their prince for ever" (Ezekiel 37:25). The second was the salvation of the individual soul, rewarded for good works, under the guidance of God: "But God will ransom my soul from the power of Sheol, for he will receive me" (Psalms 49:15). There is not much that these biblical texts tell us about the concrete features of heaven, but we can count on it being rewarding and just—not least of its many pleasures will be the chance to watch and relish the end of those wicked ones who prospered in life (Psalms 73:17).

The New Testament, meanwhile, continued the pluralism of Jewish views of the afterlife, such as a joyful existence of the individual soul who escapes hell, resurrection in the apocalyptic kingdom of God, and eternal and intimate existence with Christ. For example, the Gospel of Luke (16:19–31) tells a parable that continues the Psalmist's (73:13) theme of economic resentment (the wicked increase their wealth with ease), which is vindicated in the afterlife. There, Lazarus gets his reward by living in heaven with Abraham as the rich man who denied him food roasts in the agonies of hell. Meanwhile, in another equally famous parable, this one apocalyptic in nature, Jesus is queried by some Sadducees on the nature of the afterlife (resurrection in the kingdom of God). The question was, who among seven brothers who were married in sequence to the same woman becomes her husband after resurrection? The famous answer is that in the afterlife there is no marriage, no embodied existence in fact. Indeed, "they cannot die anymore, because they are like angels and are the children of God, being children of the resurrection" (Luke 20:36).

A bit earlier historically, in Paul's letters, heaven looks like eternal life with Christ. For example, the First Letter of Paul to the Thessalonians (4:17) states that after the resurrection, "then we who are alive, who are left, will be caught up in the clouds together with them to meet the Lord in the air; and so we will be with the Lord forever."

But as far as heavenly journeys are concerned, the New Testament is best known for the heavenly journey of John in the Book of Revelation, with its image of God sitting in human form on a throne, surrounded by a host of angels. A heavenly elder tells John that the fate of the righteous is to stand before the throne and minister to God. "They shall never again feel hunger or thirst, the sun shall not beat on them, nor any scorching heat" (Revelation 7:16). Instead, the lamb will guide them to the springs of the waters of life.

So where did the lavish and botanical visions of this heaven come from? What store of images did the Christian Church Fathers draw on when they wrote about "drinking from the waters flowing from your spring on high" or the "region of inexhaustible abundance where you feed Israel with truth for food"?[10] Even the most abstract notions of what eternal salvation looks like—and Augustine's is philosophically lofty—draws on some version of the vivid images of heaven as the delightful Elysian Fields of the Greek and Roman imagination, with their pleasurable lawns, watered by fresh springs, and sumptuous foods.[11] Heaven as the gardens of paradise continued to play an important role in medieval, Renaissance, and modern Europe. In its more literal forms, heaven was crafted for the rural and agrarian sensibility in pleasing terms: "There, lilies and roses always bloom for you, smell sweet and never wither."[12] In more sophisticated renderings, the gardens are still there, but spiritualized, as we see in Dante's "Paradise":

> *And, in the likeness of a river, saw*
> *Light flowing, from whose amber-seeming waves*
> *Flash'd up effulgence, as they glided on*
> *'Twixt banks, on either side, painted with spring,*
> *Incredible how fair.*[13]

The subject of heaven and Islam, especially in the context of martyrdom and the seventy-two virgins awaiting the *shahid* (martyr) in heaven, has inspired considerable ridicule. But the Qur'an itself most frequently reads like another desert-based scripture when it comes to its descriptions of the perfect afterlife. For instance:

> But give glad tidings to those who believe and work righteousness, that their portion is Gardens, beneath which rivers flow. Every time they are fed with fruits therefrom, they say: "Why, this is what we were fed with before," for they are given things in similitude; and they have therein companions pure [and holy]; and they abide therein [forever]. (Yusuf Ali, trans., 2.25)

There are many similarly refreshing promises throughout the Qur'an: "Verily the Companions of the Garden shall that Day have joy in all that they do; They and their associates will be in groves of [cool] shade, reclining on Thrones [of dignity]; [Every] fruit [enjoyment] will be there for them; they shall have whatever they call for" (Yusuf Ali, trans., 36:55–57).

The scriptures of India refer to heaven, or heavens, just as the Near Eastern religions do. But as I have noted already, these do not represent the highest good or the deepest joy of the religious imagination. The revealed texts, the so-called Shrutis (Vedas, Brahmanas, Upanishads) and Smritis (histories, law books) do not describe heaven with the vividness or relish of a tourist brochure. Heaven is a wonderful place, certainly, but hardly an attractive alternative to the *summum bonum* of Hindu and Buddhist striving, namely, *moksha* and *nirvana*.[14] Still, a few scattered references in the epic texts (especially in the Mahabharata) do describe a delightful place. For example, in one famous episode a good man named Mudgala earns the right to enter heaven, where beautiful nymphs live and luscious gardens with wish-fulfilling trees grow; in this place with no sorrow or pain, only pleasant and luminous experiences await the righteous denizens. However, Mudgala, on hearing that this place is temporary, turns it down.[15]

The obvious observation about heaven in every tradition is that it is wonderful, future, and conditional. Aside from these basic facts, diversity reigns in humans' different conceptions of the afterlife. There may be philosophical, political, economic, or even artistic reasons for this diversity—Dante's heaven is far more rarefied, for instance, than Pat Robertson's. But Colleen McDannell and Bernhard Lang have uncovered an important pattern within this material. There appear to be two major views on the joys that await us in heaven. The first is the joy of "eternal solitude with God alone," while the second is the joy of being "reunited with friends, spouse, children or relatives."[16]

In the Christian history of heaven, the first view, which researchers call "God-centered," appears in the New Testament, but it has prevailed mainly in the early writings of Augustine, in medieval mysticism and monasticism, among Protestant reformers, and in much of contemporary theology. The second, "human-centered" view has prevailed in the writings of such figures as Irenaeus of Lyons, the later Augustine, in much of the popular literature of the Middle Ages and Renaissance, and throughout the eighteenth and nineteenth centuries. Both visions are beatific and highly pleasurable, of course, but the pleasure is radically different. In the God-centered heaven, the self plays no role; it is virtually submerged in the ecstasy of God's own presence, and there is no consciousness of individual existence. Meanwhile, the human-centered joy is more familiar to us from our very best moments on earth—but it is magnified and extended permanently.

What are the reasons for the difference? The change in Augustine's own work sheds light on this subject, according to McDannell and Lang. As he transitioned from the solitude of the monastic life to his public responsibilities as bishop of Hippo, Augustine abandoned the abstract, Neoplatonic heaven of a mystic in favor of a social place. The life of solitude in a monastery equates with a self-less heaven in the sole presence of God, while an engaged but faith-driven social life leads to a heaven of social and even sensual joys. This is a reasonable hypothesis: The monk has already renounced his family and has systematically dismantled his social self. Why should that very same biographical self continue to exist in heaven? Or, to put the matter in hedonic terms, the monk, like the Indian philosophers, has renounced the lower forms of pleasure (novelty or replenishment and to some extent even mastery) in favor of the highest form of pleasure—exploration. Recall that novelty pleasure is tied to objects (food, wealth, sex) while mastery pleasure overcomes the dependence on novelty but tends to strengthen the sense of individual power. Only exploration pleasure is both free and self-transcending—it is pure joy.[17] Consequently, the God-centered heaven is joyful in a noncontingent and spontaneous way, while the human-centered heaven seems more preoccupied with enjoying those things that made our life pleasant while avoiding boredom.

This social hypothesis correlates with my earlier discussion of moral communities and the nature of pleasure. We saw there that moral and hyper-moral communities (of which monasteries are the strongest example) prize more advanced forms of pleasure than natural (kin-based) communities. Eating-pleasure was one key domain where such a difference played out in early Buddhism and Christianity. It seems that the difference extends to heaven. Still, even the monk who gives up family and social relations, as long as he actually believes in a future heaven, undermines the quality of his present existence. If heaven is a goal, located in the future and contingent, it can only decrease happiness in the here and now.

Why the Afterlife Makes Us Unhappy

Imagine two criminals who are sentenced for the same crime. One of them is sentenced to fifteen years but told that with good behavior he could get out in ten. The other gets ten years. Which of the two has an easier time in prison? Which would you rather be—given a choice?

Views of the afterlife that involve rewards based on judgment (scrutiny) decrease our level of happiness in life in a variety of ways, some obvious and others entirely counterintuitive. We shall begin with the obvious. The biblical texts with apocalyptic notions of the afterlife—those that describe a kingdom of God (Matthew 24:6–8) where some enjoy a wonderful existence but many perish horribly—tell us that certain signs will herald the arrival of this event. These signs are invariably negative—mostly wars and rumors of wars. Translated to the modern idiom, these harbingers would look a bit like terrorism, political instability, international conflict, corrupt government, or false Christs, such as New Age and self-help gurus—or any guru, really.

Now, when the signs are negative and you either believe or long to believe in a wonderful afterlife, you will see only negative signs everywhere. I regularly host a few young Jehovah's Witnesses in my home in Gaithersburg, Maryland. I cannot seem to persuade them how pleasant it is to sit in Tel Aviv on a winter morning, drinking espresso and munching on a real croissant, then strolling down the beachfront walkway. Unable to imagine a delightful day in Israel, they insist that horrible things are coming to that part of the world (the Holy Land they call it) and that there is not a moment to waste. They could be right, of course. Many Israelis look at Syria and Iran with grave concern, but for solutions they look to the armed forces and diplomacy, not into their own souls.

I suspect that the prisoner who received fifteen years with possible parole at ten would have the apocalyptic kind of anxiety as well. After all, his sentence is to be scrutinized, his every action observed and recorded. He is like the prisoner in Jeremy Bentham's perfect nineteenth-century prison with its "panopticon"—the structure that allows the warden to observe all his prisoners.[18] He will also have to account for himself before the parole board when the time comes. He begins to examine the warden's face, looking for signs. If the warden is having a horrible week and greets the prisoner with a nasty growl for several days in a row, the man may become alarmed and depressed. Looking over your own shoulder is hard work, and its fruit is anxiety and neurosis. The religious individual who believes in heavenly or apocalyptic reward and in God's total surveillance is like this prisoner—a paranoid who is right about his enemies.

This anxiety not only undermines the enjoyment of the present moment (and hinders performance and learning) but also raises a serious philosophical question about the effectiveness of a rewarding afterlife as an ethical and spiritual motivator. The underlying assumption of reward systems—which I shall qualify shortly—is that reward motivates. Give the dog a cookie and he'll roll over for you. If the reward is pleasurable enough, and because all of us (including our dogs) look to maximize pleasure as a matter of fact, common sense tells us that rewards should work. Is this not the reason why so many religions insist on pleasurable heavens, along with their law codes?

The problem with this commonsensical assumption is that "pleasure motivates" must apply to real, not imagined, pleasure.[19] The teacher with the gold stars gives them away every now and then. If heaven as reward only means hope for heaven or thoughts about heaven, then it is the expectation that carries the motivational load, not heaven. But given that the thoughts about heaven are accompanied by doubts, anxiety, or a sense of irrevocable contingency that can never be fully removed, the idea of heaven can no longer be purely pleasurable. At best, the idea is mixed. The only way to eliminate this dissonance is to avoid thinking about heaven, to redefine what constitutes a sign that one is doing well, or to try to create a heavenlike utopia in this life—like the Epcot Center in Disney World or other wonderful theme parks.[20]

But do rewards or incentives really motivate people, and do they make life intrinsically better? The limited benefit of rewards—a subject that ran pitched battles against behaviorism (positive and negative reinforcement) throughout the twentieth century—has recently made it into the press.[21] The school board of New York City has decided to launch a monetary reward system to improve the performance of city students in school, which presumably means achieving better test scores. Students who pass exams will get paid and can earn up to $500 per year. Some have criticized this system as one that bribes students to study, sending a lousy message. But Barry Schwartz, who is a professor of social theory and social action at Swarthmore College, wrote an op-ed piece for the *New York Times* making a strikingly different argument. Based on thirty years of research in motivation theory, he concludes that the reward system may improve the test performance of some students, but that overall the quality of the students will decline.[22] The argument is quite simple, though counterintuitive: When people have more than one reason for doing something (such as studying), the reasons do not add up but rather subtract from each other. Take the following scenario. Imagine that thirty college students who like to watch *Seinfeld* are asked to perform a small task after the next episode they watch: they are to fill out a form assessing the quality of the episode. We assume that the more they enjoy the episode, the more highly they will rate its quality. (A direct question about their enjoyment might tip them off.) Half the students are paid for this task, either with money or by being told that they will be graded and given extra credit. According to motivation theory research, the paid group will rank the episode less favorably—as a matter of rule. This principle was systematically elaborated by Mark Lepper and David Greene in 1978 and reported in their seminal work *The Hidden Costs of Reward*.[23] Edward Deci of the University of Rochester discovered the same idea by showing that when he paid a small amount of money to students who liked to solve puzzles, they lost their interest in the activity during their spare time. Over one hundred studies have confirmed these findings.[24]

The general principle can be stated simply: doing something for extrinsic reasons, such as a reward (or incentive), diminishes the motivation to do it and also diminishes the satisfaction with the task itself. The addition of a payment for watching a TV show you were already watching out of pleasure does not add another reason to enjoy it! On the contrary, it takes away from the pleasure of watching the show. Similarly, if your life is truly motivated by the thought of gaining heaven, or by the thought of pleasing God or getting praise from your rabbi, you will be less happy. The ancient Jewish moralists knew this, by the way. They instructed pupils (in Avot 1.3) to avoid working for the praise of their rabbi. Like Alfie Kohn, who wrote *Punished by Rewards: The Trouble with Gold Stars, Incentive Plans, A's, Praise, and Other Bribes* (1996), the rabbis knew what made students happy: Torah study and service of the rabbi, not the rewards.[25] Menahem Nahum of Chernobyl, the early Hasidic thinker, was explicit about this: "But the truth is that: The joy you should have in fulfilling a commandment is a true spiritual joy, something of the world-to-come . . . your joy is in the act itself, and in that way do you find joy in God."[26]

The students who get paid for watching the show may tell themselves it has become a job. Or they may become concerned with watching the show accurately, rating it correctly, or pleasing their employers. They would probably remain unaware of why their assessment of the episode of *Seinfeld* goes down because they may attribute their feelings to one thing (the show) when what now truly motivates them is entirely different (the job).

Religious people attribute their feelings to the object of their faith (God, heaven, scripture) without necessarily being aware of their own motivation (pleasure, social approval). They may not even know that they could be much happier in this life if they eliminated the notion of heavenly reward from their religion. This is really not such an absurd suggestion. In the second half of the twentieth century, several American theologians, following the philosopher Alfred North Whitehead, began to reject the idea of an afterlife, including heaven. Hans Jonas, Gordon Kaufman, Charles Hartshorne, and others argued that one did not have to be a secularist or an atheist in order to reject the concept of a heavenly afterlife—or at the very least, an afterlife that felt like a permanent Disneyland vacation. However, most people outside the divinity schools at Harvard and Chicago refuse to make the separation, at some cost to themselves.

Readers may object here and claim that polls show that people who believe in God tend to be happier (and healthier) than those who do not.[27] Assuming such polls are valid, this remains a matter of apples and oranges. It is not the belief in God that diminishes happiness, but the belief in heavenly reward for one's present actions. Those readers who believe in God and in heaven, following the academic theologians need only peel away their belief in reward.

There are two further reasons why faith in heavenly rewards diminishes happiness in this life. These two reasons, despite being elusive, are the most compelling of all. The first I shall call the "deferred gratification fallacy," and the second the "addiction to heaven syndrome."

In the popular imagination, heaven is invariably associated with pleasure. The running waters and cool breezes are like the Garden of Eden—a delightfully refreshing place (even if you are not from the Middle East, where such images are truly heavenly). And although there are no casinos or Playboy mansions in heaven, the place is all about innocent sensory pleasure (cool water, refreshing food, shade). It is thus hard to escape the impression that heaven is set up to provide us with the lasting pleasure that so eludes us in life, owing to the hard exigencies of life or, for the luckier ones, the hedonic treadmill, or habituation. If we return to the adaptive nature of pleasure discussed in a previous chapter, heaven offers replenishment (novelty) pleasure without depletion—satiation without thirst. But we also saw earlier that the Dalai Lama, Saint Ignatius, and other religious masters explain that true happiness depends not on replenishment but on escaping the adaptive cycle altogether. They teach that a purpose or goal must be set up that transcends the depletion-replenishment cycle. It can be virtue, service to the community, faith in God, and so forth. The secret to happiness is working toward a goal that takes you away from the dynamic rhythms of your personal needs. Happiness, on this view, depends on mastery or curtailment, which leads to freedom. It requires discipline, courage, and resolve. The problem with the heavenly paradise of cool water is that reward does not qualify as a purpose. If the spiritual habits that a good person establishes are aimed at deferred heavenly gratification (not sexual, as Jesus told the Sadducees, but definitely gratification), he or she will not develop the skills taught at the school for happiness. If you merely defer pleasure now for eternal pleasures later, you will not have learned to overcome the hedonic trap. This is why a good Jew must learn to actually love the Torah (Law) and why Krishna reminds Arjuna over and over again (in the Bhagavad Gita, the Hindu scripture) to avoid letting his motive be the fruit of action. Expectation of rewards is a prescription for misery.

The psychological side of lousy spiritual habits can be devastating. Acting for the sake of heaven will turn you into a spiritual addict. You will find yourself compelled to act entirely because you *want* something (reward) and not at all because you *like* it (joy). This requires an explanation.

In the common use of language, we assume that acting for the sake of pleasure means that when we enjoy doing something, like sailing on the Chesapeake, we want to do it for that reason. If you eat crab cakes and a spinach salad at a restaurant and clearly like the crab better, the next time you come to that restaurant you will probably order the crab cakes again and maybe forgo the spinach. Behaviorism has put a scientific spin on this idea—that pleasure breeds desire (wanting). However, hedonic research, particularly the work of Kent C. Berridge and his colleagues, has shown that the relation between liking and wanting is far more complex.[28] To begin with, wanting and liking depend on distinct brain circuits and neurotransmitters. By controlling dopamine levels, for example, the research can obtain states in which the subject (a lab animal) likes something without wanting it, and vice versa. Wanting may have evolved as a goal-oriented feature of our motivational system: when you want, there is some specific object attached to the feeling. Liking, on the other hand, is diffuse and lacking in such specificity; you somehow just feel good. It is possible, Berridge speculates, that wanting evolved separately as an incentive system and was later utilized by the liking system as a way to predict whether the organism would like something. Or perhaps wanting evolved as a common neural currency that allows us to choose among the different things we like, namely, food, sex, play, and so forth.

The distinction between the two systems makes a difference in terms of motivation and decision-making. Consider the following example: In the September 2003 issue of *Natural History*, Robert Sapolsky tells the heart-warming tale of Jonathan, a young male baboon, who had become smitten with the aristocratic Rebecca, a "confident young daughter of one of the highest-ranking matriarchs." All Jonathan wanted was a chance to groom her and maybe, just maybe, have her groom him. He rarely got his wish, and only once did Rebecca actually feign grooming him, in a moment of absentmindedness. That onetime pleasure was enough to keep him pursuing her. But if his goal was reproduction, what was he thinking? What mode of reasoning kept him motivated to bet so persistently against the odds? According to long-standing brain research, it is the injection of the neurotransmitter dopamine into the frontal cortex (by means of projections from the ventral tegementum) that keeps a monkey hard at work on achieving a desired goal. If the goal is pushing a lever that releases dopamine, the monkey (humans too!) will drop everything else and just continue to pull on the lever to administer the dopamine. That is the very essence of addiction. But new research has added two surprising wrinkles to this familiar fact. It turns out that the dopamine is secreted to its highest levels just *before* the pulling of the lever, when a light informs the subject to do so. And what is more, when researchers reduced the percentage of dopamine administration to only 50 percent of the times the lever was pulled, the dopamine level was actually highest! In other words, if the relation between action and consequence is only chance, the motivation to perform that act—its addictive power—is highest. The lesson, as most of those bar trawlers know, is that it's the chase that keeps them coming back for more. A Brad Pitt would get bored very quickly.

This anecdote confirms Berridge's distinction between liking and wanting. It shows that wanting, which is subserved by dopamine, is aroused by uncertainty, by chance. Gamblers, those who are honest with themselves, know that this is true: it is not the ambition to enjoy their earnings that motivates them, but the rush of the gamble, the chance. Winning $10 million will not make them less inclined to gamble.[29]

Religious writers tend to be starkly aware of the distinction between liking and wanting: Augustine was a veritable expert on the topic. Buddhist and Hindu meditators often speak about obtaining states of desireless joy. On the flip side, mystics (Christian, Muslim, and Jewish) often speak of a state of yearning for something that is always beyond reach. Vince Miller, my colleague at Georgetown, has shown how similar this yearning can be to consumerism in America.[30] The consumerist psychology, of course, is that state of constant wanting, which leads to buying and then nearly immediate dissatisfaction, which leads to more of the same. Consumerism looks like the same hedonic treadmill discussed earlier, but it is more about desire than gratification. In fact, it is the absolute dominance of wanting over liking—the spiking of dopamine just before the pulling of the lever or the signing of the credit card bill. Miller argues that objects cannot satisfy the depth of human longing—a beautiful word for wanting. Berridge would consider this the permanent separation of wanting from liking, with the former dominating.

Faith in heaven is a theological extension of such a psychology. Heaven becomes the ultimate object of wanting, like the gambler's "one last" win. It is profoundly satisfying only in the sense that it cannot disappoint until you get there. It is stimulating in that it is uncertain. It is spiritually addictive without appearing to have the negative consequences of addiction. But this is not entirely accurate: like all addictions, faith in heaven diminishes one's capacity to experience happiness now.

Skeptics will question my use of the concept of addiction to describe the condition of those who believe in heaven. For one thing, they will point out, addiction is usually generated by substances (caffeine, nicotine, alcohol, opiates, cocaine, amphetamines) or by behaviors (gambling, watching television, using the Internet, playing video games).[31] All of these produce a need that must be gratified instantly, whereas heaven is a gratification that appears to be delayed into the far future. Addictions as normally understood also have profoundly disruptive influences on life, such as struggle for control, inability to quit, denial, and harm to oneself and to one's social relationships. Addiction to heaven does not appear to be so dysfunctional: few people mortgage their homes and give away all their savings because they expect to enter heaven through the eye of a needle. Moreover, the belief in heaven aims at reinforcing socially constructive behavior—such as cooperation and obedience to legal and moral rules.

But we are not discussing heaven, strictly speaking. The subject is the future, our near-universal mental habit of focusing on the next moment rather than the present, and the pervasive expectation that the next moment should somehow fix what is wrong with the present moment, namely, the fact that we do not know how to like it. The present ennui, fear, sense of injustice, boredom, or what Krishnamurti describes as a lack of freedom requires that we depend on the next moment as an addict depends on a fix. Faith in heaven is just an extension of this need: it provides a temporary solution by means of eternity. Although comparing faith in heaven to the symptoms of addiction may be a bit of a stretch, consider this: an initial dysphoria leads to a craving for a fix, which may or may not produce a momentary euphoria. This is not only a clinical description of the phenomenology of addiction but also the way Buddha described the human condition.

From a hedonic point of view, a strong faith in heaven causes, not addiction to novelty pleasure ("instant gratification"), but addiction to mastery pleasure. Like the anorexic young woman who becomes enslaved to fasting in order to obtain her power fix, the heaven-addict must choose the actions that maximize mastery over enjoyments. That is how he gains entry to heaven. What sets him apart from the old-fashioned casino gambler is that, in the process, he also feels that he is gaining the respect and admiration of his cohorts.

Conclusion

It is possible that I have overstated the case for addiction in the case of belief in heaven—even for those who are fanatical about the afterlife. There are no studies—as far as I know—of the biology of faith in the afterlife. Still, even a cultural practice like faith in heaven has emotional and behavioral aspects, and these can only be driven by biology. Given the state of both brain and religion study, I am content to leave this point open. However, if you are playing chicken with an opponent who passionately believes in heaven and has made numerous choices throughout his life based on a calculation of a joyful and eternal afterlife, you may wish to know what truly drives him: Is it a stubborn idea? Dopamine (perhaps adrenaline) addiction? Or both? If you assume that knowing the answers to these questions makes no practical difference as far as winning or surviving such a game, you may be gambling away your own life.

And that is the challenge we face. Those who truly and passionately believe in heaven are indeed rational actors (chicken players can be rational!), but in a very specific way.[32] They do not seek to maximize their benefit (win) by taking into account the broader context (strategic planning, compromise, and so forth), but by making a clear and immediate choice (win or die, heaven or hell), which may be accompanied by strong visceral emotions. Hedonically, because this actor seeks to maximize mastery pleasure, he cannot be intimidated or induced. Just because the heaven-junkie is not rational in the usual sense, however, does not indicate that his conduct is unpredictable or unswayable. On the contrary, in some ways his behavior is more predictable and therefore more subject to influence—but not in the expected manner.

If there is a correlation between heaven and social costs and benefits—and there is—then the solution for defeating an opponent is not to undermine his faith in heaven but to increase his social costs relative to the benefits. We saw in an earlier section that the quality of heaven (God-centered versus people-centered) is correlated with social organization. Monks believe in a God-centered afterlife, and laypeople look forward to spending eternity with family, friends, and lovers. This means that heaven depends on social investment and the fruits (benefits) of such investments. The actor must improve his odds for making it to heaven by accumulating social capital in whichever type of group he lives.[33] For the one who lives in society, socially applauded actions lead to heaven. Finding yourself opposing such a person in a game of chicken, you will not persuade him that self-sacrifice is worthless, and you will probably fail to dissuade him from believing in heaven. Nor will you convince your opponent's society that heaven is just a myth or that its idea of what constitutes capital (for example, self-sacrifice and honor) is meaningless.

But you can still drive a wedge between the heaven-junkie and his society and push up the cost of being a martyr to unacceptable levels. One way to do this is discussed in the next chapter.

CHAPTER 9

The Martyr's Theater

IT HAS BEEN A WHILE since CNN aired one of these, but many of us remember them well: the video production of the suicide bomber's last will and testament. The quality is always low, but for the target audience—the parents and the home population—these are dramatic films. There are lots of props: the *kafiya* head wrap, the uniform, the polished Kalashnikov or machine gun, perhaps even the explosive belt the star will wear on his mission. There is the red, white, black, and green of the flag and in some cases an image of the Al-Aqsa mosque in the background. On a rare occasion, the mother will make an appearance as well, explaining why she is so proud of her son.

The video art of suicide terror may not impress Western audiences, but it uses powerful effects that move the right viewers. The music, often choral, is stirring. Whenever possible, the operation itself is filmed. A narrator's voice, both impassioned and authoritative, puts it all in context. Suicide terror is art, with its four great acts, the martyr's statement, the noble death (where improvisation reigns), the funeral, and the final—undoubted—denouement: a glorious entry to heaven.

Can the fact that this is all holy show business help us eliminate the culture of martyrdom? An absurd contrast comes to mind: In one famous scene from the movie *Dirty Harry*, Inspector Callahan (Clint Eastwood) is called to deal with a man who stands on the ledge of a building threatening to jump to his death. Callahan goes up in a window-cleaning crane bucket and addresses the suicidal man. But instead of trying to talk him down, Callahan asks for the man's name and address because, after the jump, he will be too "mashed" for identification. The man's despondency turns to rage, and he jumps onto the crane to attack Callahan, who knocks him out with one punch. The scene is funny because the man on the ledge intends to turn his supposed depression into a spectacle, which, to Callahan, would be a public disturbance. What the man needs is not therapy but time in lockup.

Martyrdom means many different things to the Palestinian, the Iranian, or the ancient Christian or Jew. But there is one thing that martyrdom has always been, regardless of time and place, and that is theater. Martyrdom is a carefully scripted performance designed not only for public consumption but first and foremost to satisfy an aesthetic sensibility. Whether the martyr is killing innocents or just self-immolating, it is the form of the action that counts the most. Anything else is just being a terrorist or a victim. Historically speaking, before martyrdom became theater it was literature, and before that it was spectacle. But before all of these came the ritual—the primordial source of the martyr's genre. Anyone who wishes to put an end to the culture of martyrdom must recognize it as performance art.

It may seem to readers that in calling martyrdom theatrical I am underestimating its seriousness and failing to appreciate the religious fanaticism or political despair of the suicide bomber or the divine commitment of the true religious martyr. After all, theater is "play," a ludic imitation of life, not the real deal. In this chapter, I demonstrate that it is the theatrical form of martyrdom that makes it compelling to the performers and to the audience. The pleasures of martyrdom to those who adore martyrs (being uplifted and feeling inspired, awestruck, ecstatic, and joyful) feed on the theater. The end of martyrdom appreciation, I suspect, will also have to be theatrical.

Martyrdom and Theater

THERE IS SOME AMBIGUITY in the way scholars define martyrs.[1] As most of us understand it, a martyr is someone who willingly dies for God, country, or Truth. Famous examples of this kind of death abound—especially among Christians (Bishop Polycarp was the first). Jewish history also provides stark examples: Rabbi Akiva is the most famous, but there are earlier examples in the biblical Books of Daniel and Maccabees.[2] In Islam, Al-Hallaj is probably at the top of a very long list of voluntary deaths that we regard as martyrdom. On this understanding, the victims of the Holocaust cannot be regarded as martyrs because they never had a choice but to die.

But this basic understanding is not enough. Most religious historians today think of martyrdom as something that goes beyond voluntary death on behalf of an ideal and falls into a specific literary genre. Socrates died for an idea and the zealots on Massada committed suicide, but none of these are regarded as martyrs. It is even common knowledge among experts that when a Christian was savaged by a lion in the Roman arena, the viewing public and the authorities regarded him as merely a criminal, while the lion treated him as lunch.[3] Simply dying in such a manner does not qualify as martyrdom. Indeed, it is not even clear what men like Polycarp or Akiva actually thought during their last painful moments or during the days leading up to their death. What we have come to *assume* they thought is already a product of literature—and this is the genre of the martyr. As the University of California at Berkeley scholar Boyarin puts it: "Being killed is an event, martyrdom is a literary form, a genre."[4] As such, like a comedy of manners, a Gothic romance, or a Shakespearean drama, the genre of the martyr must have its distinct elements:

- Special visions, often couched in codes
- A strong affirmation of identity or of faith (for example, the Jewish "Shema," or "I am a Christian")
- Death not as a personal choice but as a fulfillment of a general decree
- An emphasis on powerful affirmative feelings (such as love, loyalty, joy)
- An audience

The death of a stoic warrior on the battlefield is not part of this genre, either as literature or as theater. There are times when political authorities will try to turn a military casualty into a pseudo-martyr—the recent case of Corporal Pat Tillman who died by friendly fire in Afghanistan comes to mind. But as his own family insists, the military's effort to turn Tillman into a martyr was lame. The case of the contemporary Muslim *shahid* (suicide bomber) is far more interesting because it uses the props of theater, and it is the theatricality of martyrdom—its aesthetic quality—that has made it a compelling tradition in all three Western religions.

I have been using the words "theater" and "theatrical," which need to be clarified. The *Oxford English Dictionary* gives a rich definition of "theater" that in paraphrased form covers these general meanings: A. The place where performances take place, like the Brooks Atkinson Theatre on Broadway. B. A complex and stylized form of communication like the one that was first described by Aristotle in *Poetics* as "Tragedy." This style ranges from the rigid forms described in *Poetics* all the way to avant-garde and postmodern performances, including Samuel Becket and Richard Schechner who look to undermine the classical forms. C. In the final sense "theatrical" applies as an adjective to other activities, whether artistic or not, and rarely in a positive sense. For example, the soccer player who takes a dive without being touched is being theatrical.[5]

The classical (Aristotelian) understanding of theater involves a strong correspondence between the form of the performance and its subject matter. The tragedy looks to represent its subject matter (the tragic event) on the stage, to imitate it, in order to evoke powerful emotions, leading to a cathartic satisfaction at the resolution of the narrative. The performers do not have unlimited freedom in what they opt to do onstage—they are constrained by the tragic form.

The martyr's performance is theater in both of the first two senses: as a spectacle in the Roman ring it meets the first criterion, and as a stylized or structured mode of communication it meets the second. And like the classical tragedy, the subject matter and the structure of the martyr's performance are closely related.[6] The following case, one of the most familiar and well documented of all early martyrdoms, illustrates the central point of my argument.

The Passion of Perpetua

The Passion of St. Perpetua is an extraordinary historic document because it is the oldest authentic diary of a woman from ancient times and an intimate record of her time in prison, awaiting execution. The text was edited by an anonymous editor, whom most scholars believe to have been Tertullian (160–235 CE), one of the most important fathers of the Latin Church.[7]

Vibia Perpetua, born to an elite family around 180 CE, was arrested with several other members of a Christian cell, including her servant Felicitas. Perpetua was a young married woman with two children, including a nursing infant, two living parents, and two brothers. According to the *Passion*, she did not mind prison at all. In fact, she comforted her relatives and claimed that prison had become a palace for her as soon as she was allowed to nurse her infant son. She experienced several visions, which apparently reveal the Montanist (early Christian sect) elements of her faith, or at least those of the editor Tertullian.[8] These visions included the milk of sheep and being offered a morsel of curd by a white-haired Shepherd—all of this representing a variation on the wine and bread (blood and body) of the Catholic Mass.

Perpetua's father suffered more from her incarceration than any of the other relatives. A moving passage describes him coming to plead with her to compromise with the authorities. "Look upon thy mother and mother's sister; look upon thy son, who will not endure to live after thee. Forbear thy resolution."[9] Perpetua was sympathetic but resolved to stay her course. This is what she wrote about her father's sadness: "And I was grieved for my father's case because he only would not rejoice at my passion out of all my kin."[10]

Perpetua was brought before the procurator Hinarianus and commanded to perform a sacrifice on behalf of the emperors Septimus

Severus and his son Caracalla. Her father was present and begged her to think of the infant boy—but Perpetua remained unswayed. She publicly affirmed being a Christian and was sentenced, along with her brethren, to execution by beasts and gladiators in the arena. Back in prison, she suffered from swollen and painful breasts, but her father refused to deliver the infant to her. Eventually this discomfort ended. Further visions were recorded in her diary, including a pleasure garden, as promised her by her Lord, with rose trees and flowers, with four angels and with the blessed presence of previous martyrs. She also dreamed of engaging in a gladiatorial battle with an Egyptian enemy, who was the devil. In her dream she herself was a man.

Toward the end of the *Passion*, the narrator is no longer Perpetua herself. The narrator reports that on the day of the "games" (March 7, 203 CE), the victory day for the Christian martyrs, they went into the amphitheater (in Carthage) as though entering heaven. "Cheerful and bright of countenance; if they trembled at all, it was for joy, not for fear."[11] The text describes the blood baptism that took place in great detail—it was not a simple mauling, a swift execution. Nor were the martyrs passive recipients of violence. Instead, they moved around the arena, handled themselves in a very specific way, and were fully in command of their own end.[12] For example, Perpetua, in loose robes (after having first been stripped), took the trouble of covering her exposed thigh and of arranging her hair with a pin while in the arena: she did not want the spectators to think her in mourning with disheveled hair. She also did not feel the tearing of her flesh by the crazy cow (heifer) and, according to the narrator, was surprised to see the wounds. At the end, it was she who placed the sword of the novice gladiator on her own neck.

Needless to say, *The Passion of St. Perpetua* is celebrated by the Catholic Church as a record of spiritual greatness—of saintliness.

March 7, the day of the execution, is celebrated as the day of the feasts of the saints (Perpetua and Felicitas).[13]

However, the *Passion* is not just a valuable historical record or Church emblem. It is also great pulp literature. It has all the elements of a popular genre from that period, written for high-class women. According to A. D. Nock and Lacey Baldwin Smith, the Hellenistic romantic novel, which emerged in the second century, often featured a heroine who had to defy the wiles and violence of seducers and torturers.[14] She welcomed pain in order to publicly display her own purity, courage, and steadfastness. In the Christian variations on this literary genre, the heroine did not, at the end, marry her true love, but instead chose to die, often violently, in order to find true happiness in the temple of God.[15] In the *Passion* too, we read of an independent woman, contemptuous of traditional family roles, following her own passionate beliefs and demonstrating public indifference to pain and a complete freedom of will. But this literature also plays up the athletic—if one may call it that—aspects of the genre. The gladiatorial spectacle had its own rules and forms of satisfaction, to which I shall return shortly.

Perpetua is the perfect early martyr for the Church because she gives up both blood and milk—both symbols of kinship relations—in favor of an abstract idea that stands over and against two concrete institutions: the family and the state. That idea is the new society of the Christian Church. Perpetua marks the passage from what I described in chapter 5 as the kin-based social group to the voluntary community of the Church. It seems fair that it would take a woman to mark this passage because it was women who symbolized and embodied kinship.

We see this elsewhere in the classical world. According to the cultural theorist Judith Butler, following the German philosopher Hegel, Sophocles' Antigone had also once represented stark social contrasts: blood-based kinship versus the polis (political society).[16] Both Perpetua and Antigone willingly give up a husband and a son, but the Greek woman continues to cherish her parents and brother, thus still affirming the value of family against the city. In contrast, the North African woman (Perpetua) is far more radical; she rejects every conceivable political and biological relationship. Perpetua's is a revolutionary affirmation of a completely new social order. It is the moral community I discussed in earlier chapters—a voluntary association based on faith and spiritual love.

The Passion of Perpetua is theatrical in two major senses. It consists of a dramatic narrative that unfolds in the proper—enjoyable—sequence, and its denouement takes place on an actual stage, the Carthage amphitheater, although many of the other moments, especially the affirmation "I am a Christian" before the procurator, also have the feel of stage.

All of this makes one ask, what is this "play" about? I have noted that in the Greek tragedy there is a connection between the subject matter and the staging, and this appears to hold in the *Passion* as well, but what exactly is the subject matter? On a surface level, the answer is obvious. The play is about God and Christ, about evil, oppression, faith, courage, free choice, cruelty, and final reward. The structure of the narrative is the dramatic unfolding of a cosmic conflict, a deadly and earnest imitation of Christ's own drama. This much is clear and easy to outline, even if we take into account the pulp fiction elements mentioned earlier. But under the surface, there may be something still more decisive:

1. The true subject of Perpetua's theater is social identity.
2. The martyr's play, in order to be gratifying, must obey primordial and universal rules of exposition.

These two facts—the social and the hedonic—account for the central feature of the martyr's theater, namely, that it celebrates self-destruction.

Social Life as Dramatic Theater

EVERYONE KNOWS THE STORY of Jacob and Esau, the near-twins in the Book of Genesis. Jacob, the younger of the two, aided by his mother Rebecca, fools his father Isaac, who is blind with age, and wins the right of inheritance. Almost as familiar is the story of the elderly King David, who has to fend off an uprising by his son Absalom—whom the general Abner then kills. Or, to turn away from familiar biblical literature, imagine a traditional society in which a fourteen-year-old boy disobeys his father's orders or a girl refuses to marry the man her parents have chosen. In many parts of the world, this behavior would result in a major conflict, perhaps violence. I am familiar with notorious cases of "honor killings" in the Muslim and Druze communities of Israel—all it takes is a girl marrying for love and against her family's wishes. Indeed, where would world literature be without these conflicts, or every other form of social conflict?

The distinguished American anthropologist Victor Turner calls these social dramas. The philosopher Kenneth Burke thought of them more broadly as dramas of living. These dramas are not limited to ancient times or to family conflicts. Turner cites a few notorious modern examples in the Dreyfus affair, the Watergate scandal, the Boston tea party, and the U.S. embassy takeover in Tehran. It would be very easy to add contemporary examples from the Clinton and Bush presidencies—or even, more specifically, something that happened today, however trivial, in your home or workplace.

These are all dramas by virtue of sharing the structure that Aristotle identified for tragedy. They are events that unfold, or are experienced in time, according to a pattern that would have to be crafted for emotional effect if they were not already so ordered. As social dramas, Turner states, they proceed along four main steps: breach, crisis, redress, and reintegration (or acceptance of schism).[17] There is an initial event, with social implications, that causes a conflict-inducing crisis. This is met either constructively or violently, and some status quo is finally reached. The sequence applies as much to the falsely accused Dreyfus as it does to the teenager Juno in the recent hit movie. This dramatic structure does not need to be crafted, and it is extremely pervasive, for two reasons: (1) the nature of the conflict and (2) the nature of the solution in relation to human consciousness.

Conflict

The most basic tension in human social existence is between too much social control and too much freedom. Too much social control plays up collective concerns, while too much freedom is about individual wants. Cooperation as a social value competes with the natural selfishness that drives individual desires. This is a simple fact that can be described on many levels, from evolutionary principles (for animal societies too) to psychology, sociology, and even ethics.

The stresses caused by the two competing forces often cause ruptures and breach events. In many traditional societies, these breaches have a juridical quality even if what is violated is not law, properly speaking. For example, outside Varanasi, India, is a place—the tomb of Baba Bahadur Sayyid—where women who become possessed with spirits and ghosts come for exorcising.

These women, most of them young, wear ragged saris and crouch on the ground with loose hair flying about wildly as their head rotates above the shoulders like a mechanical doll that is about to break. Many of them scream or cry, frothing at the mouth as they rock back and forth, often beating their foreheads against the stones of the courtyard. Concerned family members sit nearby and patiently wait for the miraculous healing of the Muslim holy man's burial place. It is they who reveal to any visitors what instigated the crisis—invariably a conflict between tradition and the woman's personal needs. She does not wish to marry just yet, or not the man assigned to her, or she wishes to go to college instead of marrying, or she has failed to get pregnant and cannot deal with a judgmental mother-in-law. It is always a disruption of the fragile balance that society tries to maintain on the shoulders of its individual members.

The Solution

In traditional societies, the solution to these breaches includes telling the story of the event, or a similar story that bears on it. The stories at Baba Bahadur Sayyid are about an eccentric or perhaps evil aunt—the one who never married and never attained happiness—who recently died and now has taken over the body of the young victim to do her bidding. But more important than such instructive narratives are the rituals that implement a solution—just as a trial does. In fact, some of the rituals are trials, but there are also rites of divination, exorcisms, sacrifices, potlatches, and other feasts. Like a judge, and in a procedure that mimics a trial, the specialist (*ojha*, or exorcist) identifies the supernatural culprit and begins negotiations for a settlement. The terms demanded by the ghost may be two years in college, allowing the girl to wear makeup and nail polish, or other similar concessions to her individual needs in the face of the oppressive social demands on her.[18]

The ritualized solution—telling the story, identifying the culprit, and negotiating a settlement—takes on the aesthetics of drama on a stage. It appeals to the same human need for perceiving orderly events unfolding in time according to a meaningful sequence. Victor Turner got this insight from the important German philosopher Wilhelm Dilthey: the dramatic structure of the ritual literally trains human perception of past, present, and future as it imitates the structure of consciousness—our awareness of existing in time. Social dramas, experienced as suspenseful stories, acquire "meaning" in the act of telling or acting out. Not only are the women "healed" but also their relatives gain a sort of intellectual satisfaction from coming to see the essential orderliness of their own social existence. This assures them—all of us in fact—of the predictability of their acts, which depend on a reliable social context for their fulfillment. Meaning builds on this, and that makes the sacrifices we have to make for the social order more palatable. The unfolding drama is thus both ritual structure and a manner of perceiving our existence in time.

This theory echoes the way I spoke in chapter 4 about happiness: mastery over the individual adaptive process (the cycle of novelty and response) allows individuals to derive satisfaction from higher-order achievements, and these are the secret to happiness. This is not sublimation in the Freudian sense, but a developmental skill that builds on evolutionary psychology, though it is transmitted in cultural products such as music and storytelling. Here we see the mechanism at work to reduce internal social conflicts through an aesthetic experience of rituals, which either resemble or lead to theater.

The Spectacle Where the Loser Wins

If social dramas and their ritual or theatrical solutions are culture's ways of expressing and relieving social pressures, the script of the martyr is harder than most to decipher. What is the "breach," and how does martyrdom resolve it? After all, the girl who needs to be exorcised does not embrace her own suffering—so what should we make of a woman who participates in a ritual that culminates in her own violent death?

To understand the theater of the martyr, we need to look at the performance of rituals, which are more contests than trials, and try to understand when it is the loser of the contest who wins. As many scholars today know, contests in the ancient world owed much of their existence to the rituals that followed death. The passing away of a beloved relative in every society and culture triggers a range of feelings, which are always channeled by funerary and mourning rituals. However, the demise of a particularly important individual—the head of a family or clan, a wealthy property owner, a hero—is especially traumatic as a social crisis. It is this type of death that calls for the larger-scale rituals that include contests. Numerous historians of sports and theater and even classicists such as Walter Burkert, who is also a psychologist of religion, describe the rituals that attended such deaths in ancient Greece and analyze the games, which were called *agon*.[19]

The funerary rituals began with a theatrical display of grief (tearing of the hair, beating of the chest, wallowing in ashes), then proceeded to the offering of gifts to the deceased, executing sacrifices, and finishing up with a feast or banquet. In other words, the rituals moved the mourners from loss and mourning to the joys of social reintegration.[20] The games or contests were held either in connection with the sacrifices—sacrificial contests were prevalent in ancient Israel, Persia, and India as well—or later on as an independent event, linked perhaps to calendrical festivals. Such contests are described in the poetry of Homer in the *Iliad* and can be seen visually depicted on grave vessels. By the seventh century BCE, the games had become linked to hero cults; eventually they gave way to Panhellenic contests. For example, the famous games in Olympia were associated with the death of the local heroes Pelops or Oinomaos.[21] The many events—participants were no longer family but a new class of athletes—included the discus throw, chariot races and fights, jumping off chariots, sprints, wrestling, boxing, the long jump, and the javelin throw. Each event was symbolically associated with a mythical event, such as Apollo's toss of the discus.

In ancient Rome too, the most famous "athletic" events—the gladiatorial fights—were associated with the passing away of important individuals. The duels began as offerings to the shades of the dead (*manes*), enacted after the funeral. The Romans believed that they had inherited this practice from the Etruscans and the Samnites. Their own first recorded gladiatorial contest, involving three duels, was in 264 BCE. It was held by the sons of Iunius Brutus Pera at his funeral. As time went by and the wealth of Romans increased, the fights grew in size and became more lavish—despite legal efforts to curtail the slaughter.[22] This brings us closer to the theme of the martyr, the contestant who chooses to lose, or at least intentionally die, in the contest.

For indeed, the Roman games, unlike the Greek, were clearly about death. The men in the arena were marked for dying—they were either condemned criminals or captured soldiers from around the Roman world. But like the Greek games, these men wore costumes that linked them to mythical events, and the theater in which they died included choreography, props, and cheering partisan audiences.

Despite their highly elaborate forms, the Greek and Roman funerary practices, including games, demonstrate a number of basic principles that conform to Turner's social drama. There is the breach—the death of a leader or hero. This triggers a great display, a staging really, of extreme grief in the form of tearing hair, wallowing in the dirt, and beating of the chest. The next step is the sacrifice—the violent destruction and offering of an animal. At the end comes the more or less lavish banquet. So, to repeat: breach, crisis, redress, and final reintegration, all of these steps undertaken in relation to a social group that finds itself under threat of falling apart.

But unlike the relatively simple drama of a discontented young woman in India, the Greek and Roman games—and the genre of the martyr that builds on them—conceal far more elaborate social dynamics. At least three basic facts stand out here: (1) the sequence of the ritual involves an exchange of goods and services; (2) these exchanges are often violent; and (3) there is a vicarious victim, and there are assigned tasks to be accomplished in a number of steps. All of these have to be displayed in order to be effective.

Put in somewhat crude terms, this entire performance is designed to justify the division of spoils (inheritance). The surviving women need to demonstrate to the heir that they deserve continued support; the heir needs to demonstrate to the spirit of the deceased that he is worthy of inheriting the property; and the heir must show the same to his relatives and the larger social group. His "right" to inherit is contingent on both his self-control and his generosity—that is, his willingness to sacrifice for the group. What the performers are doing, in other words, is exchanging performance and wealth for social capital (acceptance of new status) in order to keep the group together.

At the center of all this mutual assurance of continued social tranquillity is the animal that must be sacrificed. It is on the victim's shoulder (or head) that anxiety, jealousy, and resentment are eliminated, or at least reduced.[23] But the exchange of the victim's life for social tranquillity must have worked to some extent, because the notion of vicarious death is extremely widespread among the religions of the world. In fact, it is the very heart of the Christian understanding of sacrifice. And if the sacrifice (and following it, the banquet with its own slaughters) does bring the community together, this shows that a third party—a ritual victim—can act to reduce disruptive impulses and build positive ones.

The reader may begin to see where this is leading: the martyr, a bit further down the road, will be the individual who chooses to act as this sacrificial victim, also for the sake of a prized social order. However, we are not ready to make that connection quite yet.

In time, both the Greek and Roman games replaced the sacrifice altogether and developed their own athletic ethos that went beyond just being the fastest or strongest. As competitions became regional and national, some athletes gained great fame and wealth. But while winning produced glory, there were other prizes—specifically, honor. This equally valuable social treasure led to a reevaluation of what constitutes victory, especially in combat. Norbert Elias tells the story of Arrhachion of Phigalia, who had twice won the Olympics in the pancration (combat). In his last fight (in 564) he was strangled to death, but as his last act he managed to break the toes of his opponent, who quit the fight. The corpse was thus declared the winner.[24] Honor, in such a case, goes to the fighter who shows the deepest commitment, but this is not simple honor. It is something more valuable and rarefied than prestige, something worth dying for. It is Truth, or Being—that which truly exists has to be negotiated, and the test of validity is the willingness to die for it. This is a difficult point to grasp for modern readers.

According to Elias, the Greek games were as violent as they were owing to the fact that the stable and impersonal monopolization of violence in the Greek city-states was still rudimentary. The repugnance against extreme violence that we feel today—in other words, conscience—only emerged later in tandem with a centralized monopoly of violence. In the absence of such authority, which we take for granted today, notions such as "right" and "truth" had to be constantly redrawn.[25] In the context of the fight, these notions do not emerge from victory (which is up to the gods) but from the manner of fighting and dying. A noble death vindicates one's conviction by attesting to its truth. There is no other criterion, or if there is, humans cannot know it in any other way. To put it starkly, our consciousness of reality is ultimately political. But this is what makes the victim powerful—that he succeeded by his performance to convince the audience of his worth as the proper emblem of their beliefs and his own. His noble death vindicates everyone's sense of reality.

This continued in the Roman games, at least according to Carlin Barton. While earlier gladiators had been slaves and captured soldiers, in time many free Romans began to fight as well. The odds of surviving any given fight varied greatly with one's skills, but even the best gladiators did not have a better chance than one in seven of surviving an entire career. It was often the master's decision whether his gladiator was to die in the ring on a given day. On rare occasions, in a splendid display of generosity, even the best of gladiators, or several of them, would be offered up by the owner to die.

Given that the role of the gladiator was lowly, debased, and doomed, why did the Romans identify so closely with their horrendous fate, and why did so many free Romans (by the end of the Republic) wish to become gladiators? According to Barton's influential study, the answer was in defying fate and cheating death by freely choosing to risk everything. The gladiator, even the slave, did not simply go into the ring. Instead, he took an oath, called *sacramentum gladiatorium*: "I shall endure being burned, bound, beaten, and slain by the sword" (*Uri, vinciri, verberari, ferroque necari patior*).[26] The oath made his obligation to fight and die contractual. In other words, his death was his own free choice, even if he was a slave to a master or to circumstances.

To return to the main theme: The contests and spectacles began as the ritualized response to the breach caused by the death of an important man. His death created a crisis of power; it stole away the center from the group. And because in archaic groups the center, with its power, was also what counted as Truth, this led to a crisis of faith. The solution, paradoxically, was also death. The noble death of a worthy victim in the arena achieved at least two major goals: it stole back from death its power of depriving humans of their beliefs—because the victim chose to die and thus put death to shame—and the surrogate victim united the spectators around the single truth of death-overcome. His performance repaired the breach of the original death and reaffirmed the center.

Of course, the arena could be a highly divisive place where spectators chose opposing fighters. Some rulers capitalized on such divisions. Similarly, the performer in the arena might have tried to usurp the meaning of the spectacle by refusing to fight and choosing to die like the best of the best. This is what Perpetua was doing, and the message was clear: I am not a frightened criminal but a willing warrior for a stronger truth—the new social order. This is not a message that the lunchtime spectators in the stands may have deciphered, but the readers of the *Passion*, the true audience, did.[27]

The Shahid Theater

Christian martyrdom had taken over the form of funerary contests to create its dramatic forms. It found its origins in death rituals and developed a theatrical disdain for death in the ring. As we saw, this willingness to accept defeat had been a sort of conquest over the archaic view of truth, power, and the social center.[28] A woman like Perpetua could never have come to define the new Christian community if she had not chosen her own violent death. But this is the Christian martyr, who honed her performances in the Greco-Roman world. How does this help us understand the contemporary *shahid*—Palestinian, Lebanese, Iraqi, or Afghani? To make the connection, we must look at Iran, or Persia, where funerals also once marked the social drama of a death that disrupted social order and where a theater of martyrdom emerged—the only significant theater in the Muslim world.

This theater, called Ta'ziyeh, has roots that reach very far back to pre-Islamic Persia—to Zoroastrianism with its mourning rituals, laments for the dead, and various storytelling performances.[29] The word itself signifies expression of sympathy, mourning, and consolation for the departure of a loved one. The theater became a type of nationalistic-religious performance staged against Arab Sunni hegemony under the Persian Safavid dynasty (1501–1735), which developed Shi'ism as the state religion. The division between Sunni and Shi'ite Islam—the former drawing on the heirs Muhammad himself appointed and the latter on his family descendants—still remains the sharpest break within the Islamic world. The Ta'ziyeh theater grew into a richly evolved artistic form in the mid-eighteenth century. It peaked during the rule of Nasseredin Shah (1848–96), who built the famous and impressive Takiyeh Dowlat (theater). The stage performances combined countless elements of religious doctrine, mythology, folklore, and ancient Persian literature. But the very core of the show marked Muharram, the period of the martyrdom of Imam Hussein, who was the son of Ali ibn Abi Talib, the cousin and son-in-law of Muhammad and the fourth caliph. It was this relationship, according to Shi'ites, that made Hussein the legitimate successor of Muhammad. The Ta'ziyeh was thus connected to the mourning rituals attending the martyrdom of Hussein and affirmed Shi'ite Islam in Persia. As such, the theater was roundly criticized by Westernized Iranian nationalists and finally was banned by the Pahlavi regime in the 1930s. It has seen a revival with the Iranian Islamic revolution and is quite popular today in Iran.

The Ta'ziyeh is a story about death and succession. It recounts the events that took place in Karbala, Iraq, in 680 CE involving the two sides of the Sunni-Shi'ite split. On one side were the elders of institutional Islam, led by Yazid—the caliph who followed Ali. On the other side were the family members of Ali, led by his two sons, Hassan and Hussein. Hussein had known that his opposition to the ruling caliph meant certain death, but he remained defiant. He wished to avoid a military confrontation, but the small group of seventy-two that he led was surrounded by the much larger force, which kept them cut off from water and food for eight days. On the ninth day, Hussein offered his companions the chance to leave, but only a few did. The drama came to a violent end on the tenth day of the lunar month of Muharram, known as Ashura (meaning "the tenth"). On that day, Hussein made his final speech, which figures in every performance of Ta'ziyeh:

> Yazid makes me choose one of the two: either I draw my sword and defend my honour and religion, or surrender to shame and humility. . . . I am obliged to choose the first way. . . . Death is

> the beginning of our joy. There is only one bridge between this world and the other world and that is death.[30]

The sad events of that day—the utter destruction of the Prophet's own family, all the men killed and decapitated—is marked every year in the month of Muharram in the Shi'ite world with mourning processions, self-beatings (a funerary act), readings of eulogies, and the performance of passion plays like the Ta'ziyeh. The mourning has a strong sacrificial-vicarious dimension: mourners believe in Hussein's intercession on their behalf toward salvation on the day of the Last Judgment.

Over the centuries, the Ta'ziyeh performances have been a magnet for numerous additional elements besides this central narrative. But the entire affair is essentially a performance of death and conquest over death, saturated with powerful emotions that range from grief and empathy to anger and victory. The play recreates the ambush, siege, and killing of Hussein and his group in Karbala. A good performance packs a strong emotional punch, as we can see from one excerpt: the moment when Qasem bids farewell to his mother:

> *When I from the saddle fall*
> *And Hussein my corpse sorrowfully haul*
> *In grief my body she'll study*
> *Her hands in my gore she'll bloody*
> *Let her not mourn and languish*
> *Nor dishevel her hair in anguish.*[31]

I began this chapter with social dramas and emphasized the death of a powerful figure—a man who occupies a central place in his group, a social magnet. Not all theater in the West took root in such a soil—there are rituals and plays that mark other social dramas—but the martyr's play did. In Muslim Persia, the theater of martyrdom, the passion, is the only show in town—it is the one stage performance that defines religion, nationalism, and communal pride. The Ta'ziyeh points back to the drama of Muhammad's death and to the succession conflict that ensued.[32] More directly, it marks the death of a martyr who chose to die for a vision of Islam that was both family-based and otherworldly. Such a martyr is the vicarious but willing victim on the altar of a radically different Muslim community. This, then, is the defining "artistic" paradigm for the Palestinian shahid—whether from Hamas or Islamic Jihad—who is inspired by Hizbollah and its Iranian-nurtured death aesthetic.

Those who are in the business of fighting suicide bombings may not care about the theatrical dimension of *shahada* (martyrdom), about the history of this theater, and certainly not about its origins in funerary rituals. For security forces, it is all about finding the heads of the organization and destroying terror operations from the top down.

But personally, I am interested in the tenacity of a cultural phenomenon that goes back three thousand years to ancient Greece and Persia. I am especially intrigued by the fact that Christianity, which has certainly inspired its share of bloodshed, has not given birth to this particular form of killing in the modern era. If Islamic martyrdom in its terrorist guise is to be counteracted in the cultural arena, it may be useful to understand why Christian martyrological theater did not take on the modern terrorizing form. In fact, it has stopped being compelling theater altogether. Or put differently, if the theater of Muslim martyrs is to lower its curtain, it is essential that a different aesthetic sensibility replace it, just as has happened in Christendom.

Comedy

By the reckoning of David Barrett and Todd Johnson, who keep a database of such things, 1990 alone saw over 200,000 Christian martyrs—and the entire twentieth century saw well over 26 million! Clearly, as the *Almanac of the Christian World* indicates, martyrdom has not passed away from the world, though few martyrs today attain the fame of a Polycarp, a Perpetua, or a John Huss. It is reasonable to ask, then, why Christian martyrdom has not been recruited for political goals: why have the Serb, South American, African, or even Chinese Christians not strapped themselves with bombs set to go off in outdoor markets? The likely doctrinal answer may be that the original Christian martyrdom, that of Jesus, set a precedent of nonviolence. The paradigm does not lend itself to a marriage of martyrdom and violent resistance. Another, perhaps more persuasive explanation is that the European Enlightenment separated religion and political action. Christians still kill for politics and still sacrifice themselves for God, but not simultaneously. Such a separation is alien to Islam, where the model of Shi'ite martyrdom, in the person of Hussein, located death on the battlefield in what was very much a political struggle.[33]

I believe that the most compelling reason for the lackluster state of martyrdom in the political arena in the West is that an aesthetic genre that has always coexisted with tragedy—the comedy—has become more satisfying or meaningful in the modern world. If the tragic martyrdom owes its origins to the crisis of death that led to ritual contests, sporting events, and theater, its sibling, the comic, came from feasts, festivals, and carnivals that carried equal theatrical potential.[34] But there is a huge difference: the contest, and later the martyr's theater, resolved the social crisis of succession by keeping the group from falling apart, while comedy made it possible for things, such as social tension, to remain unresolved and disrupted.[35]

Of course, the comic did not emerge only in Greece—ancient Indian rituals were saturated with comic and ribald contests.[36] And comedy is far from being a monolithic genre: satire, farce, parody, the comedy of manners, celebration, carnival, ribaldry, burlesque, irony, slapstick, and many other forms are distinct forms of communication and theatrical performance. But one thing they all have in common, most experts agree, is their social origin and function. From this social perspective, comedy can be divided into two types: the satiric and the populist. In the former, we laugh *at* the comic, who is the butt of humor. In populist humor, we laugh *with* the comic, who, according to Robert Torrance, is the comic hero. The juxtaposition between the two broad types has been given different terms, such as ridiculous versus ludicrous, satire versus celebration, or high social rank versus social chaos.[37] For example, we laugh *at* the heroes of Philip Roth's early books (Portnoy, Epstein), or at John Updike's Rabbit, because there is something ridiculous about their reaching too far while remaining oblivious to our derision. We laugh *with* David Sedaris (the character, not the author), who is a lisper and a gay man in the conservative South but hyper-self-conscious and entirely disruptive of the norms he invites us to question. Both of these broad types are satisfying or pleasurable, as Plato knew very well, and as he tried to explain in his dialogue *Philebus*. The first type is satisfying because it affirms the social norms, which the unwitting comic violates, incurring our derisive laughter. That is what makes him the butt of our laughter. The second type satisfies by undermining (though not destroying) the reigning values in the person of the comic hero, with whom we identify.

Which is which will vary not only according to the audience but also according to the theorist who chooses to read the character as a buffoon or a rebel. For example, is Don Quixote a delusional idiot, a madman, or in fact a truly sane man? A populist theorist who promotes a reading of the character as a comic hero will argue that Don Quixote resolves to be mad because that is the way he can consciously defy enshrined values in the manner of the comic hero.[38] In a more contemporary setting, Yossarian is not a coward at all, but the only sane character in *Catch–22*, just as the brave soldier Schweik was earlier in the twentieth century. In other words, despite its ubiquity, comedy shifts in meaning; it has to be negotiated or debated in order to settle the most basic question about its message: is it subversive or conservative? The upper classes' butt becomes the hero of the lower classes—the fool is now the witty underdog. Think, for example, of the residents of Delta House Fraternity in *Animal House* in relationship to the other college groups, or Theophilius North in Danny Huston's 1988 film *Mr. North*. Some of the great comic characters discussed by Robert Torrance exhibit this tension: Odysseus, the characters of Aristophanes, Shakespeare's Falstaff, Fielding's Tom Jones, Joyce's Leopold Bloom, Thomas Mann's Felix Krull. The tension virtually defines contemporary British comedy (recall "The Cheese Shop" routine in Monty Python's *The Meaning of Life*), with it class-conscious irreverence. Or to give a more political example, while liberals regarded President Bush as an unwitting fool whose grandiosity displayed no moderating self-awareness, supporters of the president saw in this critique the sniping of intellectual and cultural elites who did not understand the common man. The skits on *Saturday Night Live* had to negotiate their way between these two irreducible social perspectives and decide whether Bush was to be the butt or the hero.

There is a psychological component to the social function of comedy. According to Henri Bergson's influential study on laughter, individuals must evolve from the egotistical into the social. What makes the comic butt ridiculous is his failure to attain self-awareness, resulting in his laughable and often grandiose bumbling.[39] Inspector Jacques Clouseau in the early *Pink Panther* movies was precisely such a figure: a man with great dignity and high social aspirations who tries to reserve a room at a resort but can only articulate the word "rim." What makes comedy a powerful social tool, according to theorists like Susan Purdie, is that it demonstrates a failure to communicate, a lack of mastery (whether intentional or not) over language.

> An elderly Jew walked up to the window at the main post office in Minsk. "Excuse me," he began timidly, "but how often does the mail go from here to Warsaw?"
>
> "To Warsaw? Every day."
>
> The old man was silent for a moment, and then asked, "Thursdays too?"
>
> —*Anonymous*

Because the self, according to psychologists, is constituted by means of language, when comedy begins as linguistic incompetence, it ends up as potential nihilism. Such humor accounts for the "incomprehensible immigrant" film genre (*Green Card*, *Borat*, *Bread and Chocolate*) or the misunderstood silent stranger, which covers much of R. K. Naryan's literary comedy (especially *The Guide*). This certainly rings true for the humor that saturates mystical literature, especially the Indian poet Kabir and Zen masters.[40] For example, "A monk asked Ummon: 'What is Buddha?' Ummon answered him: 'Dried dung.' "

While the drama of the martyr satisfies by affirming a single socially binding idea, comedy marks the split within the social group and gives pleasure by playing up the difference. The comic does not resolve social dramas, and the clown is rarely victorious. We enjoy comedy for being festive and clownish, not for repairing anything. In Christian history, coupling the days of the saints with feasts and launching the holy season of Lent with gaudy carnivals have had the effect of taming the dramatic with the comic, at least to some extent. I suspect this juxtaposition between the dramatic and the comic has cultivated an aesthetic balance in Christian cultures. The gravity and solemnity of the high rituals were thus lovingly mocked and abrogated through the levity of the carnivals beginning already in the High Middle Ages.[41] This is why Christians can tolerate, however grudgingly, Monty Python's *Life of Brian* or why Americans laugh at Mark Twain's quip: "This is a Christian country. Why, so is hell." This does not mean that Christianity had become silly and nonviolent, but that the dramatic appeal of sacred self-destruction has been moderated by other, equally satisfying (and far more self-critical) factors. The aesthetic of martyrdom, in other words, is no longer sufficiently satisfying to overcome competing genres that could easily transform the political martyr into a simple quixotic figure. We have become too self-conscious and too self-appraising to fall into such a trap. We have learned to question the martyr as a person, and that makes him lose his vicarious power.

This is not unique to Christianity or Judaism, which may be even more famous for its uses of comic genres. One of the most famous figures in Indo-Islamic civilization in early modern times, the poet Kabir, was largely a religious clown. He was a fourteenth- and fifteenth-century weaver from Varanasi, India, and a member of a caste of converts to Islam that occupied a low spot in the social hierarchy of India's holiest city. Kabir was a mystic and poet, a man who was both a Muslim and a Hindu and who wrote sharp satirical poems about Hindus and Muslims alike.

The Hindu says Ram is the beloved
The Turk says Rahim.
Then they kill each other.[42]

Naturally, he got into trouble with both communities. A popular legend reports that Kabir and Sikander Lodi, the Sultan ruler, had an encounter once in Varanasi that almost led to violence, but the story was too comical for that. The Brahmins and the Muslim leaders reported Kabir to Sikander, saying that he had abandoned the customs of the Muslims and broken the rules of the Hindus against untouchables. He had scorned the sacred bathing places and even the holy scriptures of Hinduism. The sultan summoned the lowly weaver and demanded a show of respect, but Kabir refused to bow before him. Facing the charge that he had abandoned the traditional religion of both Hindus and Muslims, Kabir replied: "The Hindus and the Turks are the ones who will fall into hell. The kazis and the mullahs are clumsy fools. I have found salvation through *bhakti* [devotion]. I have sung about the virtues of Ram through the guru's grace."[43]

Both Hindus and Muslims were enraged by Kabir's disrespect. They demanded that the sultan stone the infidel, but instead Sikander had Kabir chained and thrown into the Ganges. Miraculously, the chains fell off as soon as Kabir hit the water, and he floated away to safety. Next the sultan threw the weaver into a burning house, but the flames became as cool as water. Finally, a wild elephant was set on Kabir, but as that legend would have it, God (Hari) took the form of a lion, which frightened the elephant. With this, the sultan gave up.

This legendary tradition, which refused to have the social critic martyred, also celebrated the farcical elements of his life. Kabir was so shy that whenever guests came to his house, he ran away.

The more famous he became, the more outrageous his conduct became. To turn people away, he would walk the streets of Varanasi holding hands with prostitutes, pretending to be drunk. His poetry was not only satirical—at its best, like the words of Zen, it also violated the rules of language and logic:

The instrument is still,
Its string snapped.
What can the poor thing do?
Its player's no longer there.[44]

Such a comic is much more likely to be a mystic than a martyr, and Kabir was no martyr, but his comedy also displays the power to incite passionate social responses and to unmake the self by undermining language. It is possible that much of this literature was politically motivated (against Islamic hegemony in northern India). Still, as comedy, it is fitting that the tradition that celebrated these sorts of episodes refused to engage in the pathos of martyrdom.

However, it may be fair to ask how effective comedy truly is in political resistance or in bringing about social changes outside of legend and hagiography. Furthermore, can poking fun at one's enemies do anything more than strengthen their resolve? Comedy may reflect a high level of tolerance for internal social contradictions, but does the use of comedy promote freedom? Can the tail wag the dog? One recent case suggests an answer.

In the mid-1940s, the Ku Klux Klan was still a large and active organization in the American South. Stetson Kennedy, a young Florida writer and folklorist, infiltrated the organization in order to study it. According to Steven D. Levitt and Stephen J. Dubner's account in *Freakonomics*, Kennedy's plan to disrupt the KKK worked brilliantly because it was based on making information widely available. The most carefully guarded secrets of the Klan were exposed: passages from their bible oddly called the Kloran, the secret roles and titles of high officials, code words and names, and various messages and commands. But it was not enough that the information was revealed. Kennedy passed it on to the producers of *Adventures of Superman*, a radio show that was enormously popular with kids. After fighting against Hitler and Mussolini, Superman now took on the KKK, and clinging to his cape, so to speak, were tens of millions of little boys. Not only were Superman and all the boys fighting the Klan, but they also seemed to know everything about it. This was too much for the members, as Kennedy noted; according to one exasperated Klan member, the kids "knew all our secret passwords and everything. I never felt so ridiculous in all my life! Suppose my own kid finds my Klan robe some day?"[45]

There is some controversy over the extent of Kennedy's own work in undermining the KKK, but none over what made it work. Exposing the rituals on national radio was just simply funny and silly, both to the kids who ridiculed Superman's enemies and to the general listening public. This was too much for a serious and sanctimonious organization. An individual who becomes aware that the heroic persona he has adopted has become the butt of humor—a target not of hatred or violence but of pervasive ridicule—will attain the level of self-awareness that Bergson discusses in his analysis of laughter. When the hero becomes a comic butt, his posture loses viability. The answer to the question about the power of comedy and humor is that it can be devastating. One can hardly imagine politics without it, going back to the Greeks and Romans through the British satirists (Daniel Defoe, Jonathan Swift, and the many politicians, like Benjamin Disraeli and Winston Churchill, who were notorious for their biting wit) and the Americans (Mark Twain, Sinclair Lewis, H. L. Mencken). Where would American politics be today without Jon Stewart, Stephen Colbert, and *Saturday Night Live*? Or, on the other side, Ann Coulter?

But most theorists agree that humor has to come from within the group. Anti-Semitic cartoons may make Nazis or Iranians laugh, but they have no genuine effect on the object of derision. The Danish Muhammad cartoons incited resistance; even the hilarious spoof on the seventy-two virgins in paradise by Steve Martin in *The New Yorker* would be useless as a tool for bringing about social or religious reform in Muslim countries. Charlie Chaplin had no effect on Hitler or any other Nazi. Satire directed against foreigners is too blunt, broadly aimed, and often counterproductive. The best Jewish humor in the modern era—when Jewish humor as we know it was born—has been directed at Jewish characters (the Rothschild rich man, the Hershele fool, the rabbi's wife). For that matter, the best Arab political satire is not directed at Americans or Israelis but at other Arabs—the Egyptians against the Tunisians, the Jordanians against the Iraqis.

Could Arab comedy undermine the culture of the shahid? Is the culture humoristic enough to slay such a sacred cow (to mix in a Hindu metaphor)? As it turns out, although Arab and Muslim countries have not developed a *theatrical* tradition of comedy like what we have in the West, humor has played an enormous role in both Arabic and Islamic literature, dating back to the Qur'an itself. The history of Arab literature shows a far more prolific interest in humor than the Jews ever developed. According to Franz Rosenthal, dozens of theoretical writers, including grammarians and physicians, either wrote books that explained humor or engaged in a variety of humorous exercises themselves. A full bibliography would occupy several pages with names such as "Books on Buffoons" or "the Book on Jokes and Pleasantries."[46]

Furthermore, the tradition of laughter goes all the way back, not only to the Qur'an but also to the Prophet himself, who was known for his sense of humor and promotion of humor: "From hour to hour, refresh your hearts for if they tire, they will lose their insight."[47] One of Muhammad's closest companions, Uthman, was even more famous for his practical jokes and for tolerating jokes directed at himself. One of the most famous professional comedians in Arab history was his jester—Ash'ab the Greedy (d. 771). Once approached by his beloved asking for a ring by which to remember him, he responded that she might just as reliably remember him for *not* giving her a ring. Just as famous is the much later Hoja Nasr-ad-din (Nasreddin), whose humorous adventures took place in Turkey and were both socially satirical and self-lampooning. In one story, Hoja boasts of cutting off the hand of an enemy on the battlefield. When asked if he had then cut the man's head off, he confessed that he did not, for someone else had already done so hours earlier. Similar adventures are also attributed to the character Juha, who is even more famous in the West.[48]

Arab humor is still alive and well, in a wide variety of flavors. Even the Palestinians joke about their leadership; according to Sharif Kanaana, an ethnologist at Bir Zeit University, Arafat was a common target, and Abu Mazn is now. Here is one Arafat joke: "I want a Palestinian state," Arafat says to God, who wants to fulfill a wish for him. God hems and haws. "It will not happen in your lifetime, Arafat." "I want Jerusalem." "Nor will this happen in your lifetime, Arafat." "Then at least I want to be as good-looking as George Clooney." "Arafat!" says God. "That won't even happen in my lifetime."[49]

The problem, then, is not the absence of Arab or even Muslim humor, but its application to the most sacred values. There are hundreds of jokes about Moses and Jesus—sometimes both are mocked in the same joke—and they are told by Jews and Christians alike.

In one of these jokes, Jesus, playing golf, is compared unfavorably to Arnold Palmer. True humor is unrestrained and potentially liberating. The few traditions about Muhammad's sense of humor do not mitigate the extreme caution that attends Islamic joking about sacred matters. And that may point to the source of the problem.

Comedy is an intellectual pleasure, as even Islamic scholars have noted over the centuries. It satisfies not just by provoking deep feelings but by triggering anticipation, amazement, and recognition.[50] To be sophisticated, comedy requires the suspension of self-importance, which is a sort of intellectual error. In contrast, tragedy—especially heroic or martyr tragedy—is profoundly emotional. It satisfies by causing grief, a sense of moral injury, and fantasies of revenge. It is far more "primitive" from the perspective of social and psychological adaptation.

I believe that the most effective *cultural* way of undermining the pleasure of heroic and martyr tragedy is through comedy. Forget about arguing against the noble cause or about the calculus of cost to benefit. The target of a psychological campaign should be the individual who thinks—pretentious fool that he is—that he can presume to be like the Imam Hussein and that his own battlefield is none other than Karbala. Can you imagine a Christian who acted as though he were Jesus? A man who enters, let us say, the National Cathedral's bookshop and tosses out the cash machine and during his trial for disorderly conduct and property damage claims to be the Son of God throwing out the money-changers from the Temple? This patently absurd, if not pathetic, scenario is the stuff of comedy. One who models his actions on a literal reading of the holy scriptures risks unflattering pity, while the artist who sings the praises of such a fool risks derision. Even Mel Gibson only managed to avoid turning his *Passion* into schmaltz by using sequences of gut-wrenching violence (and Aramaic).

Once a given theme has been exposed to the intellectual joys of comedy, it is hard to recover the pathos that first drove it.[51] The video productions of shahids on their way to suicide may be frightening to victims of terror, and they may be uplifting to Palestinians who view them, though not as much as to the Shi'ites of southern Lebanon, from whence this culture migrated. Potentially—and I must emphasize that word—shahids may also become the objects of Arab satire. The characters starring in these productions could be viewed as pretentious and misguided rather than uplifting, entirely lacking in self-awareness rather than saintly. If they came to be exposed in a different light, even by those who want a free Palestine, it is possible that the culture of the shahid would go the same way as the culture of the would-be Christ or the KKK.

It is important to keep in mind that comedy is a cultural response, and a long-term one at that. It belongs in a larger arsenal of weapons for combating the culture of suicide, such as arts, sciences, and sports. The idea is to saturate the lives of would-be martyrs, not with wealth, but with alternative forms of prestigious pleasures. Because martyrdom is a unique form of entertainment, to put it crudely, comedy is the single most effective counterform.

Conclusion

SINCE 9/11, HUNDREDS OF BOOKS have been written on religion and violence, including suicide terrorism and the so-called global jihad. The prevailing assumption among religion scholars—as opposed to popular writers—is not that religion is intrinsically violent, but that it can somehow turn bad. Religion can be distorted, co-opted, or simply misunderstood in the name of aggressive agendas. At worst, one may suggest that religion contributes to separate group identities, or that it promotes absolute claims to truth and thus exacerbates preexisting conflicts. If that is the case, religion is no worse than nationalism, capitalism (or communism), or any other secular justification for human aggression.

Here we have taken a sharply different tack. I have argued that there is something unique about the religious life, an emotion or affect that is so basic and so compelling that it makes being religious almost irresistible. It is an attractive feeling, and this is precisely what makes religion so dangerous. I have argued that, above all other institutions in human history, religion has produced the most effective technology for manufacturing happiness out of the raw material of pleasure. By virtue of its unique psychological methods and moral authority, religion—and this is true for all the major religions of the world—can discipline its members to renounce simple pleasures and reach for higher-level enjoyments. The reason for such a strange task, and the reason religions still thrive, is that this is the secret to reining in individual impulses for the sake of social values. The raison d'être of religious hedonics, as I call this technology, is the group. Religion makes groups cohere, and it does this by means of spiritual happiness. This is precisely what makes religion dangerous at the brink of a catastrophe such as a nuclear showdown.

The finest and most refined product of religious hedonics is spiritual love. This is the love that radiates from God and the love that humans offer God in return. It is also the love that is refracted from the spiritual master—a mystic or saint—and draws passionate followers to him. This love, I have argued, is a form of pleasure, and it behaves like pleasure. By implication, then, if an economic psychologist who relies on hedonics can measure and predict consumerist behavior, a religion psychologist ought to be able to do the same for those who act out of such spiritual love. I hope this is the case.

Unfortunately, in some situations refined religious joys will drive actors to choose self-destruction over defeat or even compromise. Martyrs or members of spiritual groups, following the dictates of supreme and lasting joy, will decide that self-annihilation is more appealing than a compromised life. They may actually feel no rage, display no aggression, and even love their enemy. But they will choose some form of immolation. As Nietzsche noted, there is something profoundly aggressive about such joy: did Saint Perpetua not condemn her infant to life as an orphan? Even worse, religious groups that hold together around the master's love may welcome the unmaking of the entire social order in order to rebuild it in their own mold—if not in this world, then the next. If this odd form of behavior, which is as archaic as the roots of religion and as contemporary as suicide bombers, is driven by pleasure, it can be diagnosed by scientists and stopped by other forms of pleasure.

This book takes that first stab at a hedonic science of religion, religious feelings, and religious motivation. If the basics are correct, and if the theory is sound, this book may give researchers a few tools for devising methods to neutralize religious self-destructiveness. The readers have discovered methods, some of them surprising and counterintuitive, for talking down a martyr or influencing the decisions of those who seek an apocalypse. I say "surprising" because these methods include both laughter and supporting the Iranian mullahs against opponents of the regime.

ACKNOWLEDGMENTS

I would like to thank many people for the conversations we have had over the years on the subject of religion and self-destruction. The ideas in this book have grown and improved thanks to such feedback, although almost none of the friends and colleagues listed here have read the manuscript—all its errors are entirely my own. My thanks go to Carl Ernst, Ian Lustick, Jerrold Post, Fathali Mogghadam, Keith Payne, Lise Morje Howard, Maria Fanis, Sherifa Zuhur, Tom Scheber, Michelle Van Cleave, Doug Shaw, Clare Wilde, Terry Reynolds, Elizabeth McKeown, Vince Miller, Theresa Sanders, and, of course, Bud Ruf. I would also like to thank the members of the Ohio State University Program in Comparative Studies, both faculty and graduate students. I am particularly indebted to Tom Banchoff and the Berkley Center for material support, to Paul Heck for his patience and insights, to Jennifer Hansen more than ever, and finally, to Eric Brandt for his wise counsel.

NOTES

Chapter 1: Religious Self-Destructiveness and Nuclear Deterrence

1. *Jerusalem Post*, August 10, 2001.
2. Associated Press photographs of this event taken by Nasser Ishtayeh on September 23, 2001, can be seen online at: www.gamla.org.il/english/feature/sbarro.htm/.
3. An example of such an analysis is Linda M. Pitcher, "The Divine Impatience: Ritual, Narrative, and Symbolization in the Practice of Martyrdom in Palestine," *Medical Anthropology Quarterly* 12 (1, 1998): 8–30.
4. Ami Pedahzur, *Suicide Terrorism* (Cambridge: Polity, 2005); Michael Taarnby, "Profiling Islamic Suicide Terrorists: A Research Report for the Danish Ministry of Justice," submitted November 27, 2003. Taarnby discusses several examples, including Nizar Trabelsi, an Al Qaeda terrorist who had been a successful soccer player in the German Bundesliga. A perspective derived from detailed interviews with women bombers who survived is available (in Hebrew) in Anat Barko, *The Path to the Garden of Eden* (Tel Aviv: Yediot Ahronot, 2004).
5. Leila Hudson, "Coming of Age in Occupied Palestine: Engendering the Intifada," in Balaghi Shiva and Muge Gocek, eds., *Reconstructing Gender in the Middle East: Tradition, Identity, and Power* (New York: Columbia University Press, 1995).
6. See also the discussion on fig wasps in Richard Dawkins, *Climbing Mount Improbable* (New York: W. W. Norton, 1996), ch. 10.

7. David Sloan Wilson, *Darwin's Cathedral: Evolution, Religion, and the Nature of Society* (Chicago: University of Chicago Press, 2002); see also David Sloan Wilson and Elliott Sober, "The Fall and Rise and Fall and Rise and Fall and Rise of Altruism in Evolutionary Biology," in Stephen G. Post, Lynn G. Underwood, Jeffrey P. Schloss, and William B. Hurlbut, eds., *Altruism and Altruistic Love: Science, Philosophy, and Religion in Dialogue* (New York: Oxford University Press, 2001).
8. The neurological work to back up these grand judgments is still in preliminary stages but is promising indeed. See, for instance, Jorge Moll et al., "The Neural Correlates of Moral Sensitivity: A Functional Magnetic Resonance Imaging Investigation of Basic and Moral Emotions," *Journal of Neuroscience* 22 (7, 2002): 2730–36.
9. Not all philosophers are put off by the biological challenge to free will; see the work of Daniel Dennett, *Freedom Evolves* (New York: Viking, 2003). For a contrasting perspective, see John Searle, *Freedom and Neurobiology: Reflections on Free Will, Language, and Political Power* (New York: Columbia University Press, 2007).
10. Anne Marie Oliver and Paul F. Steinberg, *The Road to Martyr's Square: A Journey into the World of the Suicide Bomber* (New York: Oxford University Press, 2005), p. 140.
11. Oliver and Steinberg, *The Road to Martyr's Square*, p. 136.
12. Martin Amis, interview in *Times of London*, October 23, 2002; see also the comments by the political scientist Francis Fukuyama in "Paglia and Fukuyama on Cho, Masculinity, and Sex," *National Review Online*, April 25, 2007, available at: http://phibetacons.nationalreview.com/post/?q=MTc2NmU4Y2IyNzJhNTRiYWYyYjM3ZGNhMTcwMDNlZWQ=. The connection between suicide terrorism and religion is not universally accepted among mainstream scholars either. The most outstanding example of the rejection of such a connection is Robert Pape, *Dying to Win: The Strategic Logic of Suicide Terrorism* (New York: Random House, 2005).
13. Quoted in Shari Goldstein Klein, Dolev Gotlieb, and Shahar Shoshana, "The Operational and Emotional Code of the Suicide Terrorist: Implications for Psychological Warfare," in Hagai Golan and Sha'ul Shai, eds., *A Ticking Bomb: Encountering Suicide Terror*, my translation (Tel-Aviv: Ma'arachot, 2006), p. 79.
14. Hector N. Qirko and Scott Atran, "Fictive Kin and Suicide Terrorism," *Science* 2 (304, April 2004): 49–51.
15. Adolph Tobena and Scott Atran, "Individual Factors in Suicide Terrorism," *Science* 2 (304, April 2004): 47–49.
16. "LTTE Admits Killing Rajiv Gandhi," *Financial Express*, June 27, 2006.
17. Peter Schalk, "The Revival of Martyr Cults Among Ilavar," *Temenos* 33 (1997): 151–90, available at: www.tamilnation.org/ideology/Schalk01.htm;

see also Ami Pedahzur, *Suicide Terrorism* (Cambridge: Polity, 2005), ch. 4.

18. Itzhak ben Israel, "Fighting Suicide Terrorism: The Israeli Case," in Golan and Shai, *A Ticking Bomb*, p. 26.
19. Keith B. Payne, testimony before U.S. House Armed Services Committee, Subcommittee on Strategic Forces, July 18, 2007; Wahied Wahdat-Hagh, "Basij—The Revolutionary People's Militia of Iran," *Inquiry and Analysis Series* 262 (February 1, 2006).
20. Jonathan Rosenblum, "Imposters in Hassidic Garb," *Jerusalem Post*, December 19, 2006; see also the December 29 interview with Yisrael Hirsch, *Jerusalem Post.*
21. Aviezer Ravitsky, *Messianism, Zionism, and Jewish Religious Radicalism* (Chicago: University of Chicago Press, 1996), especially ch. 2; for greater theological detail, see Shaul Magid, *Hasidism on the Margin: Reconciliation, Antinomianism, and Messianism in Izbica Radzin Hasidism* (Madison: University of Wisconsin Press, 2003); Jerome R. Mintz, *Hasidic People: A Place in the New World* (Cambridge, Mass.: Harvard University Press, 1992). See also Hella Winston, *Unchosen: The Hidden Lives of Hasidic Rebels* (Boston: Beacon Press, 2005).
22. *Harijan* magazine, November 26, 1938.
23. Nahum N. Glatzer and Paul Mendes-Flohr, eds., *The Letters of Martin Buber: A Life of Dialogue* (New York: Schocken Books, 1994).
24. *Hindustan Times*, August 8, 1947; see also *Collected Works of Mahatma Gandhi*, vol. 89, p. 11, and vol. 87, pp. 394–95.
25. Erik Erikson, *Gandhi's Truth: On the Origins of Militant Nonviolence* (New York: W. W. Norton, 1993), pp. 242–43.
26. M. K. Gandhi, *Gandhi: An Autobiography: The Story of My Experiments with Truth* (Boston: Beacon Press, 1957), pp. 342–44.
27. Erikson, *Gandhi's Truth*, pp. 242–43.
28. A recent book on cognitive dissonance (a term coined by Leon Festinger) explores this in a variety of ways: Carol Tavris and Elliot Aronson, *Mistakes Were Made (but Not by Me): Why We Justify Foolish Beliefs, Bad Decisions, and Hurtful Acts* (New York: Harcourt, 2007).
29. Rodney Stark and Roger Finke, *Acts of Faith: Explaining the Human Side of Religion* (Berkeley: University of California Press, 2000).
30. For a wonderful exploration of such questions, see Tim Harford, *The Undercover Economist* (New York: Random House, 2007), ch. 7.
31. The founding text of game theory is John von Neumann and Oskar Morgenstern, *Theory of Games and Economic Behavior* (Princeton, N.J.: Princeton University Press, 1944); for recent aspects of game theory, see Martin J. Osborne, *An Introduction to Game Theory* (New York: Oxford University Press, 2004).

32. Richard Sosis and Candace Alcorta, "Signaling, Solidarity, and the Sacred: The Evolution of Religious Behavior," *Evolutionary Anthropology* 12 (2003): 264–74. I return to this topic in chapter 7, where I discuss the theatrics of martyrdom.
33. One unusual version of the prisoner's dilemma can be seen in the first chapter of Ian McEwan's *Enduring Love.* That game scenario could be called "Who Will Let Go of the Hot Air Balloon First?"
34. Described in Kaushik Basu, "The Traveler's Dilemma," *Scientific American* (June 2007). Psychologists and economists know that "rationality" is a loose concept and that people decide on their interests in emotional and chaotic ways. See, for example: Dan Ariely, *Predictably Irrational: The Hidden Forces That Shape Our Decisions* (New York: HarperCollins, 2008) and Paul Ormerod, *Butterfly Economics: A New General Theory of Social and Economic Behavior* (New York: Basic Books, 2000).
35. Basu has challenged the reliability of human rationality in several publications, including "On the Nonexistence of a Rationality Definition for Extensive Games," *International Journal of Game Theory* 19 (1990): 33–44.
36. Richard Dawkins, *The Selfish Gene* (New York: Oxford University Press, 1977). Dawkins does not deny the existence of natural altruism. He seeks to explain it, such as in the case of male fig wasps, with selfish gene theory. See also his *Climbing Mount Improbable* (New York: W. W. Norton, 1996), pp. 308–26.
37. Paul C. Stern, "Nationalism as Reconstructed Altruism," in *Political Psychology* 17 (3, September 1996): 569–72.
38. Amartya Sen, *Rationality and Freedom* (Cambridge, Mass.: Belknap Press of Harvard University Press, 2002), for example, p. 159.
39. Blaise Pascal, *Pensées*, trans. A. J. Krailsheimer (London: Penguin, 1966), fragment 148. For a modern argument, see Carolyn R. Morillo, *Contingent Creatures: A Reward Event Theory of Motivation and Value* (Lanham, Md.: Littlefield Adams Books, 1995). As we shall see throughout the book, we can avoid the seeming absurdity of psychological hedonism by defining pleasure in a biological/evolutionary manner.

Chapter 2: The Mysteries of Pleasure

1. Michel Cabanac, "Pleasure: The Common Currency," *Journal of Theoretical Biology* 155 (1992): 173–200. See also Michel Cabanac, Jacqueline Guillaume, Marta Balasko, and Adriana Fleury, "Pleasure in Decision-Making Situations," *BMC Psychiatry* 2 (7, 2002): available at: www.biomedcentral.com/1471-244X/2/7; and Helen Phillips, "The Pleasure Seekers," *New Scientist* 2416 (October 2003).

2. Anthony J. Steinbock, *Phenomenology and Mysticism: The Verticality of Religious Experience* (Bloomington: Indiana University Press, 2007).
3. Pleasure is one of the "great topics" of philosophy going back to Plato's *Philebus* and *Timaeus* and Aristotle's *Nicomachean Ethics* (book 10). There is no point in attempting even a schematic bibliography. I can, however, recommend a few general contemporary works: Leonard D. Katz, "Pleasure," in Edward N. Zalta, ed., *Stanford Encyclopedia of Philosophy Online*, available at: plato.stanford.edu; J. C. B. Gosling, *Pleasure and Desire* (Oxford: Clarendon Press, 1969); Fred Feldman, *Pleasure and the Good Life: Concerning the Nature, Varieties, and Plausibility of Hedonism* (Oxford: Clarendon Press, 2004); Daniel Kahneman, Ed Diener, and Norbert Schwarz, eds., *Well-being: The Foundations of Hedonic Psychology* (New York: Russell Sage Foundation, 1999). Additional works are cited in context.
4. Ludwig Wittgenstein said: "Joy surely designates an inward thing. No. Joy designates nothing. Neither any inward nor any outward thing"; see *Zettel* (Oxford: Basil Blackwell, 1967), no. 487. See also Gilbert Ryle, *The Concept of Mind* (1949; reprint, Chicago: University of Chicago Press, 1984), p. 109. For a refutation, see Irwin Goldstein, "Intersubjective Properties by Which We Specifiy Pain, Pleasure, and Other Kinds of Mental States," *Philosophy* 75 (2000): 89–104.
5. Katz, "Pleasure," p. 7.
6. Epicurus, *Principle Doctrines* VIII, quoted in John M. Cooper, *Reason and Emotion: Essays in Ancient Moral Psychology and Ethical Theory* (Princeton, N.J.: Princeton University Press, 1999), p. 495, n. 15.
7. This follows the prescription of the Hedonist Aristipuss, who is quoted in Feldman, *Pleasure and the Good Life*, p. 32.
8. John Locke, *An Essay Concerning Human Understanding*, 2 vols. (Oxford: Clarendon Press, 1894), p. 303.
9. Jeremy Bentham, *An Introduction to the Principles of Morals and Legislation* (1789; reprint, Oxford: Clarendon Press, 1996), p. 11.
10. Bottom-up or empirical theories of pleasure (and happiness, which is the balance of pleasure to displeasure) require that we be able to quantify our enjoyment. Daniel Kahneman has been a leading figure in this research program; see Kahneman et al., *Well-being*, for an example of the measurement of "instant utility." For an alternative calculus of pleasure, see Allen Parducci, *Happiness, Pleasure, and Judgment: The Contextual Theory and Its Applications* (New York: Lawrence Erlbaum, 1995), ch. 2. Other scales are available; see R. P. Snaith, M. Hamilton, S. Morley, A. Humayan, D. Hargreaves, and P. Trigwell, "A Scale for the Assessment of Hedonic Tone: The Snaith-Hamilton Pleasure Scale," *British Journal of Psychiatry* 167 (1995): 99–103.

11. Marquis de Sade, *The 120 Days of Sodom and Other Writings* (New York: Grove Press, 1966), pp. 450–51.
12. *Nicomachean Ethics*, 1369b33–1370al, quoted in J. C. B. Gosling and C. C. W. Taylor, *The Greeks on Pleasure* (Oxford: Clarendon Press, 1982), p. 196.
13. "Each kind of being," wrote Aristotle, "again, seems to have its proper pleasure, as it has its proper function"; *Nicomachean Ethics*, trans. F. H. Peters (New York: Barnes and Noble, 2005), p. 230.
14. Saint Augustine, *Confessions*, trans. Henry Chadwick (Oxford: Oxford University Press, 1998), 10.31.43.
15. Saint Augustine, *Confessions*, 10.31.44.
16. Saint Augustine, *Confessions*, 10.23.33.
17. Rick Warren, *The Purpose Driven Life* (Grand Rapids, Mich.: Zondervan, 2002), p. 73.
18. What I call "Greek philosopher" or top-down pleasure is still highly regarded among contemporary philosophers. See, for example, Bennett Helm, *Emotional Reason: Deliberation, Motivation, and the Nature of Value* (New York: Cambridge University Press, 2001); Daniel Nettle, *Happiness: The Science Behind Your Smile* (New York: Oxford University Press, 2005); Norton Nelkin, "Reconsidering Pain," *Philosophical Psychology* 7 (1994): 325–43.
19. In Nettle's (*Happiness*, p. 4) words, it's about "striving for the goals that evolution has built into us."
20. Victor S. Johnston, "The Origin and Function of Pleasure," in James A. Russell, ed., *Pleasure* (East Sussex, U.K.: Psychology Press, 2003), pp. 167–80.
21. Johnston, "The Origin and Function of Pleasure," in Russell, ed., *Pleasure*, p. 173.
22. Richard D. Alexander, *The Biology of Moral Systems* (New York: Aldine de Gruyter, 1987); M. Cabanac, C. Pouliot, and J. Everett, "Pleasure as a Sign of Efficacy of Mental Activity," *European Psychologist* 2 (1997): 226–34; D. M. Warburton, "The Function of Pleasure," in N. Sherwood Chichester, ed., *The Functions of Pleasure* (Hoboken, N.J.: John Wiley and Sons, 1996), pp. 1–10.
23. Jaak Panksepp, *Affective Neuroscience: The Foundation of Human and Animal Emotions* (New York: Oxford University Press, 1998).
24. Charles W. Fox, Derek A. Roff, and Daphne J. Fairbairne, eds., *Evolutionary Ecology: Concepts and Case Studies* (New York: Oxford University Press, 2001); Denise Dellarosa Cummins and Collin Allen, *The Evolution of Mind* (New York: Oxford University Press, 1998). There are different ways of sorting out the types of pleasure imposed by biological and ecological factors; see, for example, Michael Kubovy, "On the Pleasures of the Mind," in Kahneman et al., *Well-being*, pp. 134–54.

25. Panksepp, *Affective Neuroscience*, pp. 164–76.
26. Peter J. Richardson and Robert Boyd, *Not by Genes Alone: How Culture Transformed Human Evolution* (Chicago: University of Chicago Press, 2005). The relationship between contemporary psychological traits and evolutionary environments is not universally accepted; see, for example, David J. Buller, "Four Fallacies of Pop Evolutionary Psychology," *Scientific American* (January 2009): 74–81.
27. Gregory Berns, *Satisfaction: Sensation Seeking, Novelty, and the Science of Finding True Fulfillment* (New York: Henry Holt, 2005).
28. Paul Rozin, "Preadaptation and the Puzzles and Properties of Pleasure," in Kahneman et al., *Well-being*, pp. 109–33.
29. Jared Diamond, *Why Is Sex Fun? The Evolution of Human Sexuality* (New York: Basic Books, 1997); Helen Fisher, *Why We Love: The Nature and Chemistry of Romantic Love* (New York: Henry Holt, 2004).
30. Rozin, "Preadaptation." See also literature on encephalization (brain size) and evolutionary psychology.
31. On the implications of biological adaptation for cultural and religious facts, see Ralph Wendell Burhoe, "Pleasure and Reason as Adaptations to Nature's Requirements," *Zygon* 17 (2, December 2005): 113–31.

Chapter 3: The Varieties of Religious Pleasure

1. Leon Festinger, *A Theory of Cognitive Dissonance* (Stanford, Calif.: Stanford University Press, 1956), pp. 30–31. Festinger's work has generated a tremendous amount of secondary literature. One of the best of these works, Michael S. Gazzaniga, *The Mind's Past* (Berkeley: University of California Press, 2000), describes how the brain (invisibly) shapes the world around us.
2. Carol Tavris and Elliot Aronson, *Mistakes Were Made (but Not by Me): Why We Justify Foolish Beliefs, Bad Decisions, and Hurtful Acts* (New York: Harcourt, 2007). E. E. Evans-Pritchard, *Witchcraft, Oracles, and Magic Among the Azande* (Oxford: Oxford University Press, 1937).
3. See Benedict Carey, "Denial Makes the World Go Round," *New York Times*, November 20, 2007, with references to the work of Michael McCullough, Peter H. Kim, Donald L. Ferrin, and other researchers who study the evolutionary (social) benefits of lies, self-deception, and other vices. See also Alina Tugend, "The Many Errors in Thinking About Mistakes," *New York Times*, November 24, 2007.
4. The concept of the false messiah in Judaism goes beyond the realm of simply mistaken perception of such a figure and into the domain of deception, hysteria, and self-destruction. On Sabbetai Tzvi, see Gershom Scholem, *Sabbatai Zevi: The Mystical Messiah: 1626–1676* (London: Routledge Kegan Paul, 1973).

5. The idea that rites of passage increase social solidarity has been one of the foundations of sociology and social anthropology from Émile Durkheim's *The Elementary Forms of the Religious Life* (New York: Free Press, 1995) to Victor Turner's *The Forest of Symbols* (Ithaca, N.Y.: Cornell University Press, 1967). More recent theorists, such as Marvin Harris, Michael Harner, and Leslie White, have focused on the relationship of rituals to ecological factors. Finally, evolutionary theories have looked at rites of passage as costly signals that the youth undergoing them will be altruistic and not freeloaders. See, for example, Richard Sosis, "Why Aren't We All Hutterites? Costly Signaling Theory and Religious Behavior," *Human Nature* 14 (2, 2003): 91–127.
6. The theory of encephelization has been influential (though controversial) in both physical and cultural anthropology and, recently, even in sociology; see Alexandra Maryanski and Jonathan H. Turner, *The Social Cage: Human Nature and the Evolution of Society* (Stanford, Calif.: Stanford University Press, 1992). For a critique, see Pierre L. van den Berghe's review of *The Social Cage*, *Contemporary Sociology* 23 (1, January 1994): 94–96; Thomas Plummer, "Flaked Stones and Old Bones: Biological and Cultural Evolution at the Dawn of Technology," *American Journal of Physical Anthropology*, suppl. 39 (2004): 118–64. A more detailed examination of encephelization, particularly with reference to the work of Paul Rozin, can be seen in chapter 4 in the discussion of food tastes.
7. The best critical summary of all such positions up until the 1960s remains E. E. Evans-Pritchard, *Theories of Primitive Religion* (Oxford: Clarendon Press, 1965).
8. Wilfred Cantwell Smith, *The Meaning and End of Religion: A New Approach to the Religious Traditions of Mankind* (New York: Macmillan, 1963). Smith's position remains influential today; see Talal Asad, "Reading a Modern Classic: W. C. Smith's *The Meaning and End of Religion*," *History of Religions* 40 (2001): 205–22; Bruce Lincoln, *Holy Terrors: Thinking About Religion After September 11* (Chicago: University of Chicago Press, 2002).
9. Influenced by a Protestant worldview, Western definitions of religion favor the belief component over the social or institutional. Compare, for example, Clifford Geertz, *The Interpretation of Cultures: Selected Essays* (New York: Basic Books, 1973), and Talal Asad, *Genealogies of Religion: Discipline and Reasons of Power in Christianity and Islam* (Baltimore: Johns Hopkins University Press, 1993).
10. Ronald Grimes, *Research in Ritual Studies* (Metuchen, N.J.: Scarecrow Press, 1982); Michael Dietler and Brian Hayden, *Feasts* (Washington, D.C.: Smithsonian Press, 2001).
11. On the difference between types of pleasure and domains of pleasure, see chapter 2. All the major religions of the world are rife with references to

pleasures and joys. Even the greatest warriors against the senses were deep connoisseurs of joy and happiness—men such as Augustine, Plotinus, John of the Cross, Rumi, al-Hallaj, Shankara, Sri Caitanya, Vivekananda, and C. S. Lewis.

12. See this chapter, note 5.
13. Ellen Handler Spitz, "Reflections on Psychoanalysis and Aesthetic Pleasure: Looking and Longing," in Robert A. Glick and Stanley Bone, eds., *Pleasure Beyond the Pleasure Principle* (New Haven, Conn.: Yale University Press, 1990), pp. 221–38.
14. On the mystical text Zohar, which is full of references to the joys of the Sabbath, see Daniel Chanan Matt, trans., *Zohar: The Book of Enlightenment* (Mahwah, N.J.: Paulist Press, 1983). See in particular references to the "secret of Shabbath"—for example, "Her face shines with a light from beyond" (p. 132).
15. Arthur Green, trans., *Menachem Nahum of Chernobyl: Upright Practices, The Light of the Eyes* (Mahwah, N.J.: Paulist Press, 1982), pp. 249–50.
16. Erla Zwingle, "Pamplona, No Bull," *Smithsonian* 37 (4, July 2006); see also Carrie B. Douglass, *Bulls, Bullfighting, and Spanish Identity* (Tucson: University of Arizona Press, 1997).
17. Zwingle, "Pamplona, No Bull," p. 94. For more on this ludic theme and its connection with the Holi in India (discussed in the next section), see Mikhail Bakhtin, *Rabelais and His World* (Bloomington: Indiana University Press, 1984).
18. Jerome R. Mintz, *Carnival Song and Society: Gossip, Sexuality, and Creativity in Andalusia* (Oxford: Berg, 1997); Roberto Da Matta, *Carnivals, Rogues, and Heroes: An Interpretation of the Brazilian Dilemma* (South Bend, Ind.: Notre Dame University Press, 1991). See also Ernest Hemingway, *Death in the Afternoon* (New York: Scribner, 1937), p. 213.
19. Robert McMahon, *Augustine's Prayerful Ascent: An Essay on the Literary Form of the Confessions* (Athens: University of Georgia Press, 1989), p. 97.
20. I discuss the union of festivals with passion plays and commemorations of martyr days in chapter 9.
21. Douglass, *Bulls, Bullfighting, and Spanish Identity*, p. 125.
22. McKim Marriott, "The Feast of Love," in Milton Singer, ed, *Krishna: Myths, Rites, and Attitudes* (Chicago: University of Chicago Press, 1966), pp. 200–12; see also John Stratton Hawley, *At Play with Krishna: Pilgrimage Dramas from Brindavan* (Princeton, N.J.: Princeton University Press, 1981).
23. The Holi in Brindavan is not unique. There are other carnivalesque festivals elsewhere in India, such as the Ram-lila in Varanasi and the singing of Radha-Krishna bhajans in Madras (Chennai), as described by Milton Singer in *The Radha-Krishna Bhajans of Madras City* (Chicago: University of Chicago Press, 1963).

24. Saint Teresa of Ávila, *The Interior Castle*, trans. Mirabai Starr (New York: Riverhead, 2003); Teresa of Ávila, *The Book of My Life*, trans. Mirabai Starr (Boston: New Seed Books, 2007).
25. Saint Teresa of Ávila, *The Interior Castle*, p. 98.
26. Saint Teresa of Ávila, *The Interior Castle*, p. 89.
27. Saint Teresa of Ávila, *The Interior Castle*, p. 167.
28. Steven T. Katz, ed., *Mysticism and Religious Traditions* (New York: Oxford University Press, 1983). The argument made by Katz and his supporters over several publications is that what appears to be the ineffable essence of all mystical experience is false. Mystics experience what their traditions prepare them to experience. The case of Teresa supports this thesis; see Mary Frohlich, *The Intersubjectivity of the Mystic: A Study of Teresa of Ávila's Interior Castle* (Atlanta: Scholars Press, 1993).
29. The four-volume edition was published by Princeton University Press in 1982. A one-volume edition is also available: *The Passion of Al-Hallaj: Mystic and Martyr of Islam* (Princeton, N.J.: Princeton University Press, 1994).
30. *The Passion* (1994), pp. 284–85.
31. Peter D. Kramer, *Listening to Prozac* (New York: Penguin, 1997), p. 264; see also Dan W. Brock, "The Use of Drugs for Pleasure: Some Philosophical Issues," in Thomas H. Murray, Willard Gaylin, and Ruth Macklin, eds., *Feeling Good and Doing Better: Ethics and Nontherapeutic Drug Use* (Totowa, N.J.: Humana Press, 1984).
32. Daniel Kahneman and Amos Tversky, "Prospect Theory: An Analysis of Decision Under Risk," *Econometrica* 47 (1979): 263–91. Loss aversion is a specific aspect of a larger phenomenon called negativity bias. See Jonathan Haidt, *The Happiness Hypothesis: Finding Modern Truth in Ancient Wisdom* (New York: Basic Books, 2006), pp. 28–31.
33. See chapter 5, notes 46, 47.
34. There is a more technical (and precise) way of putting this: "We expect costly in-group requirements to be more prevalent in communities characterized by high potential gains from collective action, low genetic relatedness and high intergroup mobility"; Richard Sosis and Candace Alcorta, "Signaling, Solidarity, and the Sacred: The Evolution of Religious Behavior," *Evolutionary Anthropology* 12 (2003): 264–74. See chapter 5 for a more detailed discussion.
35. Careful researchers of the evolution of religion try to maintain a strict separation between statements about God (or gods and goddesses) and human response to objects of faith, though it is very tempting to cross the line. See, for example, Rodney Stark and Roger Finke, *Acts of Faith: Explaining the Human Side of Religion* (Berkeley: University of California Press, 2000).

36. Cabanac has been a prolific and important experimental researcher of pleasure. For a similar study, see Michel Cabanac, Jacqueline Guillaume, Marta Balasko, and Adriana Fleury, "Pleasure in Decision-Making Situations," *BMC Psychiatry* 2 (7, 2002), available at: www.biomedcentral.com/1471-244X/2/7.

Chapter 4: The School for Happiness

1. The idea that religious morality and virtues lead to happiness has been very familiar to theologians throughout history. See for example, William C. Mattison, *Introducing Moral Theology: True Happiness and the Virtues* (Grand Rapids, Mich.: Brazos Press, 2008). What is exceptional is the notion that happiness is, in fact, a form of pleasure. The subject of happiness has gained a good bit of attention in recent years. See, for example, Darrin M. McMahon, *Happiness: A History* (New York: Atlantic Monthly Press, 2006); Daniel Gilbert, *Stumbling on Happiness: Think You Know What Makes You Happy?* (New York: Knopf, 2006); Gregory Berns, *Satisfaction: Sensation Seeking, Novelty, and the Science of Finding True Fulfillment* (New York: Henry Holt, 2006). See also the *New York Times* (Jan. 7, 2007) interview with Todd Kashdan ("Happiness 101"), who teaches a course on well-being at George Mason University. Martin Seligman of the University of Pennsylvania is generally acknowledged as one of the founders of the new field of happiness studies.
2. Steven Johnson, *Emergence: The Connected Life of Ants, Brains, Cities, and Software* (New York: Simon & Schuster, 2002), p. 19. A recent application of emergence theory to religion—God as the sum total of emergent complexity in nature—can be seen in Stuart Kauffman, *Reinventing the Sacred: A New View of Science, Reason, and Religion* (New York: Basic Books, 2008).
3. Richard Dawkins, *The Blind Watchmaker: Why the Evidence of Evolution Reveals a Universe Without Design* (New York: W. W. Norton, 1996), pp. 47–50.
4. R. Keith Sawyer, *Social Emergence: Societies as Complex Systems* (Cambridge: Cambridge University Press, 2005), ch. 4.
5. D. M. Warburton, "The Function of Pleasure," in N. Sherwood Chichester, ed., *The Functions of Pleasure* (Hoboken, N.J.: John Wiley and Sons, 1996), pp. 1–10; Victor S. Johnston, "The Origin and Function of Pleasure," in James A. Russell, ed., *Pleasure* (East Sussex, U.K.: Psychology Press, 2003), pp. 167–80.
6. Compare this with the bar patron who swings at the first suggestion of an insult; see Stanley I. Greenspan and Stuart G. Shanker, *The First Idea: How Symbols, Language, and Intelligence Evolved from Our Primate Ancestors to Modern Humans* (Cambridge, Mass.: Da Capo Press, 2004), p. 24.

7. Other scholars say this as well. The best example is Stark and Finke, *Acts of Faith*.
8. See chapter 3 for a discussion of separability (or the "Prozac effect") in Peter D. Kramer, *Listening to Prozac* (New York: Penguin, 1997).
9. See Paul Rozin, "Development in the Food Domain," *Developmental Psychology* 26 (4, 1990): 555–62; Michael Pollan, *The Omnivore's Dilemma: A Natural History of Four Meals* (New York: Penguin, 2006).
10. Isak Dinesen, *Babette's Feast and Other Anecdotes of Destiny* (New York: Vintage, 1988). The idea that high cuisine requires the same mastery (by the chef and the diner) as music and visual art has been familiar to researchers and philosophers for a long time; see Carolyn Korsmeyer, "Delightful, Delicious, Disgusting," *Journal of Aesthetics and Art Criticism* 60 (3, Summer 2002), for a full list. The notion that, in contrast, good taste and religious values must conflict, which is at the heart of the short story, is also extremely familiar. But it can be overstated, to the detriment of religious connoisseurs of various stripes, not least the Trappist monks in Latroun and elsewhere; see Frank B. Brown, *Good Taste, Bad Taste, and Christian Taste* (New York: Oxford University Press, 2000). Dinesen herself may be mocking this notion, which is what makes her story so comical; see Ann Gossman, "Sacramental Imagery in Two Stories by Dinesen," *Wisconsin Studies in Contemporary Literature* (1963): 319–26.
11. Paul Rozin, "Preadaptation and the Puzzles and Properties of Pleasure," in Daniel Kahneman, Ed Diener, and Norbert Schwarz, eds., *Well-being: The Foundations of Hedonic Psychology* (New York: Russell Sage Foundation, 1999), pp. 109–33; Paul Rozin and Deborah Schiller, "The Nature and Acquisition of a Preference for Chili Pepper by Humans," *Motivation and Emotion* 4 (1980): 77–101.
12. R. L. Solomon, "The Opponent-Process Theory of Acquired Motivation: The Costs of Pleasure and the Benefits of Pain," *American Psychologist* 35 (8, 1980): 691–712; see also Daniel Berlyne, *Aesthetics and Psychology* (New York: Appleton-Century-Crofts, 1971).
13. Greenspan and Shanker, *The First Idea*, pp. 32–33.
14. Christopher Butler, *Pleasure and the Arts* (Oxford: Clarendon Press, 2004), p. 85.
15. Bruno Bettelheim, *The Uses of Enchantment: The Meaning and Importance of Fairy Tales* (New York: Vintage, 1989), pp. 41–45.
16. Bettelheim, *The Uses of Enchantment*, p. 42.
17. Marina Warner, *From the Beast to the Blonde: On Fairy Tales and Their Tellers* (New York: Farrar, Straus and Giroux, 1995), p. 23.
18. The topic of storytelling, or narrative formation, within developmental psychology is enormous. Among the most prominent voices in this field

is that of Jerome Bruner; see his book *Acts of Meaning* (Cambridge, Mass.: Harvard University Press, 1990).

I shall return to this when I discuss the theater of the martyr (chapter 9). See Victor Turner on social dramas as theater: *From Ritual to Theatre: The Human Seriousness of Play* (New York: Performing Arts Journal Publications, 1982), p. 69.

19. Meir Sternberg, *The Poetics of Biblical Narrative: Ideological Literature and the Drama of Reading* (Bloomington: Indiana University Press, 1985), pp. 309–20.
20. Reading enjoyment has not attracted many fans in the Modern Language Association in recent decades, but there have been a few prominent exceptions; see, for example, Frank Kermode, ed., *Pleasure and Change: The Aesthetics of Canon* (Oxford: Oxford University Press, 2004). But there are other, more esoteric, enjoyments to be discovered in every form of art; see Christopher Butler, *Pleasure and the Arts: Enjoying Literature, Painting, and Music* (New York: Oxford University Press, 2004); see also Stanley Fish, "Think Again," *New York Times*, January 13, 2008, on the pleasures of "unpacking literary texts," that is, uncovering nonobvious meanings. This domain of literary appreciation is explored in detail by David S. Miall and Don Kuiken, "A Feeling for Fiction: Becoming What We Behold," *Poetics* 30 (2002): 221–41. Or, as William H. Gass puts it, in the long run you will enjoy a game of chess only against an equal or superior player; see *A Temple of Texts* (New York: Knopf, 2006), pp. 7–8; see also John Sutherland, *How to Read a Novel: A User's Guide* (New York: St. Martin's Press, 2007).
21. For telling examples, see Ulrich Baer, *Spectral Evidence: The Photography of Trauma* (Cambridge, Mass.: MIT Press, 2002).
22. Daniel Anderson, who helped create Nickelodeon's *Blue's Clues*, is quoted in Malcolm Gladwell, *The Tipping Point: How Little Things Can Make a Big Difference* (NewYork: Little, Brown, 2002), p. 126. Of course, the youngsters learn by noticing, understanding, and enjoying new things in each viewing.
23. The subject of religion as storytelling is enormous. One of the most poignant examples (and my favorite) is the way Hasidic storytellers turned Lurianic Kabbalah into stories that were both redemptive and fun; see Joseph Dan, *The Hasidic Story: Its History and Development* (Jerusalem: Keter, 1990). Martin Buber turned the story of the Hasidim themselves into entertaining literature (*Tales of the Hasidim*).
24. Plato, *Laws*, II 659(e)1–5. For a discussion of the meaning of this quote, see Luc Brisson, *Plato the Myth Maker* (Chicago: University of Chicago Press, 1998), p. 77.

25. Leonard B. Meyer, *Emotion and Meaning in Music* (Chicago: University of Chicago Press, 1956); see also Judith A. Peraino, *Listening to the Sirens: Musical Technologies of Queer Identity from Homer to Hedwig* (Berkeley: University of California Press, 2006); Susan K. Langer, *Feeling and Form: A Theory of Art Developed from Philosophy in a New Key* (New York: Charles Scribner's Sons, 1953). On the neurological underpinnings of expectation as a cognitive tool, see Wolfram Schultz's work on monkeys: "Predictive Reward Signal of Dopamine Neurons," *Journal of Neurophysiology* 80 (1998): 1–27.
26. David Huron, *Sweet Anticipation: Music and the Psychology of Expectation* (Cambridge, Mass.: MIT Press, 2006); see also Daniel J. Levitin's enjoyable discussion, focusing on music's activation of the pleasure centers, in *This Is Your Brain on Music: The Science of a Human Obsession* (New York: Dutton, 2006); Laird Addis, *Of Mind and Music* (Ithaca, N.Y.: Cornell University Press, 1999). Biological theories of aesthetics run into a variety of difficulties, as illustrated in John C. Marshall's review in *Nature* 341 (1989) of Ingo Rentschler, Barbara Herzberger, and David Epstein, eds., *Beauty and the Brain: Biological Aspects of Aesthetics* (Boston: Birkhauser, 1988). However, neuro-aesthetics is a growing and increasingly sophisticated field. See, for example, Hideaki Kawabata and Semir Zeki, "Neural Correlates of Beauty," *Journal of Neurophysiology* 91 (2004): 1699–1705; Semir Zeki, "Art and the Brain," *Daedalus* 127 (2, 1998): 71–103.
27. If the music progresses in a way that plays up the rhythmical structures, like the Indian raga, while the melodic line remains constant, Western audiences tend to lose interest. On the mathematics and enjoyment of music, see Dmitri Tymoczko, "The Geometry of Musical Chords," *Science* 313 (5783, 2006): 72–74.
28. See also Laurence Perrine, *Story and Structure* (New York: Harcourt Brace, 1983).
29. The reader is invited to give this theory a test. Listen to Verdi's "Va, Pensiero" and note your reaction when the choir sings the two lines beginning with "Apra d'or" and "Le memorie." There are several versions of this piece on YouTube.
30. Huron, *Sweet Anticipation*, p. 315.
31. Huron, *Sweet Anticipation*, p. 315.
32. Warner, *From the Beast*, p. xx.
33. Greenspan and Shanker, *The First Idea*, p. 35.
34. Dan Messinger, Alan Fogel, and Laurie Dickson, "A Dynamic Systems Approach to Infant Facial Action," in James A. Russell and Jose-Miguel Fernandez-Dols, eds., *New Directions in the Study of Facial Expression* (New York: Cambridge University Press, 1997).

35. Sheri Goldstein Klein, Dolev Gotleib, and Shahar Shoshana, "The Operational Code and the Emotional Code of the Suicide Terrorist's Attachment to Terror Organizations," in Hagai Golan and Sha'ul Shai, eds., *A Ticking Bomb: Encountering Suicide Terror* (Tel-Aviv: Ma'arachot, 2006), pp. 73–93.
36. This topic is discussed in great detail by Michel Foucault in his book *The Use of Pleasure: The History of Sexuality*, vol. 2 (New York: Vintage Books, 1990). The discussion is based on classical Greek material in which, for example, a variety of training forms (*askesis*) were designed to enhance the well-being of the soul by training the body's response to pleasure (p. 74). Foucault also quotes Socrates (in the *Gorgias*) on the essential relation between moderation (which is hedonic discipline), beauty, and happiness (pp. 89–90). The subject of religion and discipline has been very widely discussed. Two important recent articles are directly relevant: Michael E. McCullough and Brian L. B. Willoughby, "Religion, Self-Control, and Self-Regulation: Associations, Explanations, and Implications," *Psychological Bulletin* (2009, forthcoming); Ayelet Fishbach, Ronald S. Friedman, and Arie W. Kruglanski, "Leading Us Not Unto Temptation: Momentary Allurements Elicit Overriding Goal Activation," *Journal of Personality and Social Psychology* 84 (2, February 2003): 296–309.
37. Dalai Lama and Howard C. Cutler, *The Art of Happiness: A Handbook for Living* (New York: Riverhead, 1998), p. 13.
38. Some researchers have argued that this return to steady-state is an indication that happiness is genetically rooted, but this does not mean that happiness cannot be optimized. See Jonathan Haidt, *The Happiness Hypothesis: Finding Modern Truth in Ancient Wisdom* (New York: Basic Books, 2006).
39. The difficulty of gauging our own level of happiness is the premise of Daniel Gilbert's *Stumbling on Happiness* and the many studies it cites.
40. This insight is widely shared throughout the religious literature of South Asia. See, for example, Bhagavad Gita 2.38: "Hold pleasure and pain, profit and loss, victory and defeat to be the same" (Zaehner).
41. Dalai Lama and Cutler, *The Art of Happiness*, p. 42.
42. Anthony Mottola, trans., *The Spiritual Exercises of Saint Ignatius* (New York: Doubleday, 1989), p. 102.
43. Mottola, trans., *The Spiritual Exercises of Saint Ignatius*, p. 47.
44. Mottola, trans., *The Spiritual Exercises of Saint Ignatius*, p. 78.

Chapter 5: Disgust and Desire: Why We Sacrifice for the Group

1. Alan Forey, *Military Orders and Crusades* (Aldershot, Hampshire, U.K.: Variorum, 1994), ch. 2.

2. Gerard Chaliand and Arnaud Blin, *The History of Terrorism: From Antiquity to Al Qaeda* (Berkeley: University of California Press, 2007), pp. 55–78.
3. Ami Pedahzur, *Suicide Terrorism* (Cambridge, Mass.: Polity, 2005), pp. 43–69.
4. Pedahzur, *Suicide Terrorism*, pp. 43–69; see also Hagai Golan and Sha'ul Shai, eds., *A Ticking Bomb: Encountering Suicide Terror* (Tel-Aviv: Ma'arachot, 2006).
5. Rodney Stark, *Acts of Faith: Explaining the Human Side of Religion* (Berkeley: University of California Press, 2000). The subject of renunciation and martyrdom among the middle classes and the elites is immense; see Alan B. Krueger, *What Makes a Terrorist: Economics and the Roots of Terrorism* (Princeton, N.J.: Princeton University Press, 2007); James A. Francis, *Subversive Virtue: Asceticism and Authority in the Second-Century Pagan World* (University Park: Pennsylvania State University Press, 1995).
6. Richard Dawkins, *The God Delusion* (New York: Houghton Mifflin, 2006).
7. Marshall Sahlins, *The Use and Abuse of Biology* (Ann Arbor: University of Michigan Press, 1977). This was only one response (but the most prominent) in the wave of intense and angry responses to Edward O. Wilson, *Sociobiology: The New Synthesis* (Cambridge, Mass.: Harvard University Press, 1975).
8. Paul Rozin, "Towards a Psychology of Food and Eating: From Motivation to Module to Model to Marker, Morality, Meaning, and Metaphor," *Current Directions in Psychological Science* 5 (1, 1996): 24; Sharman Apt Russell, *Hunger: An Unnatural History* (New York: Basic Books, 2005).
9. Robert Axelrod, *The Evolution of Cooperation* (New York: Basic Books, 2006).
10. Axelrod, *The Evolution of Cooperation*, pp. 27–54.
11. Axelrod, *The Evolution of Cooperation*, pp. 65–68. Clusters that gain traction are said to "invade" a population of noncooperators. The cluster must be small enough to gain traction and not too widely dispersed or recognizable through some type of marking.
12. Robert Boyd and Peter J. Richerson, *Culture and the Evolutionary Process* (Chicago: University of Chicago Press, 1985); Robert Trivers, *Social Evolution* (Menlo Park, Calif.: Benjamin/Cummings, 1985).
13. Elliott Sober and David Sloan Wilson, *Unto Others: The Evolution and Psychology of Unselfish Behavior* (Cambridge, Mass.: Harvard University Press, 1998); see also David Sloan Wilson, *Darwin's Cathedral: Evolution, Religion, and the Nature of Society* (Chicago: University of Chicago Press, 2003).
14. Food can literally act like a flag; see Abbebe Kifleyesus, "Muslims and Meals: The Social and Symbolic Function of Foods in Changing Socioeconomic Environments," *Africa: Journal of the International African Institute* 72 (2, 2002): 245–76.

15. A very substantial sociological and anthropological literature examines the role of food in social structure, going back most prominently to Marcell Mauss, the French scholar and a student of Durkheim; see Anna Meigs, "Food as a Cultural Construction," in Carole Counihan and Penny Van Esterik, eds., *Food and Culture: A Reader* (New York: Routledge, 1997), 102; Igor de Garine and G. A. Harrison, eds., *Coping with Uncertainty in Food Supply* (New York: Oxford University Press, 1988); Mark Conner and Christopher J. Armitage, *The Social Psychology of Food* (Philadelphia: Open University Press, 2002), especially introduction and ch. 2; William Alex McIntosh, *Sociologies of Food and Nutrition* (New York: Plenum Press, 1996); Helen Macbeth and Jeremy MacClancy, *Researching Food Habits* (New York: Berghahn, 2004); Jack Goody, *Cooking, Cuisine, and Class: A Study in Comparative Sociology* (Cambridge: Cambridge University Press, 1982). On the relation between food quality and social stratification, see Marijke van der Veen, "When Is Food a Luxury?" *World Archaeology* 34 (3, 2003): 405–27; Hortense Powdermaker, "Feasts in New Ireland: The Social Function of Eating," *American Anthropologist* 34 (new series) (2, 1932): 236–47.
16. Mary Douglas, *Purity and Danger: An Analysis of the Concepts of Pollution and Taboo* (London: Routledge Classics, 2002).
17. E. P. Sanders, *Judaism: Practice and Belief, 63 BCE–66 CE* (London: SCM Press; Philadelphia: Trinity Press International, 1992), p. 435.
18. On food in Hinduism, see R. S. Khare, *The Eternal Food: Gastronomic Ideas and Experiences of Hindus and Buddhists* (Albany: State University of New York Press, 1992); K. T. Achaya, *Indian Food: A Historical Companion* (Delhi: Oxford India Paperbacks, 1994). The enjoyment of food, expressed metaphorically, is discussed in V. Aklujkar, "Sharing the Divine Feast: On the Food Imagery of Marathi Sant Poetry," Khare, *The Eternal Food*, pp. 98–99. The important devotional text Caitanya Caritamrita (trans. E. C. Dimock, Harvard Oriental Series) also displays loving attention to gastronomic metaphors; see 3.10.104–24; 3.2.107; 3.18.99–103.
19. See Wendy Doniger O'Flaherty, *Karma and Rebirth in Classical Indian Traditions* (Berkeley: University of California Press, 1980); Ariel Glucklich, "Karma and Pollution in Hindu Dharma: Distinguishing Law from Nature," *Contributions to Indian Sociology* 18 (1, 1984): 25–43.
20. The concepts of purity and pollution, which are central to Judaism, Islam, and Hinduism, are gradually disappearing from current scholarship. The most influential twentieth-century book on the subject in India was sociological: Louis Dumont, *Homo Hierarchicus: The Caste System and Its Implications* (Chicago: University of Chicago Press, 1980). For a qualified and nuanced view of where the concept stands today "on the ground," so to speak, see Sarah Lamb, *White Saris and Sweet Mangoes: Aging, Gender, and the Body in North India* (Berkeley: University of California Press, 2000).

21. Edward J. Lawler and Shane R. Thye, "Bringing Emotions into Social Exchange Theory," *Annual Review of Sociology* 25 (1999): 217–44; Steve Bruce, *Politics and Religion* (Cambridge: Polity Press, 2003).
22. On the cultural role of natural and acquired disgust, especially its recruitment for group marking, see Paul Rozin, Jonathan Haidt, Clark R. McCauley, and Sumio Imada, "Disgust: The Cultural Evolution of a Food-Based Emotion," in Helen M. Macbeth, ed., *Food Preferences and Taste: Continuity and Change* (Oxford: Berghahn, 1997), pp. 65–82. In the hands of Claude Lévi-Strauss, attitude toward food could serve as the very foundation of culture; see, for instance, *The Raw and the Cooked*, trans. John and Doreen Weightman (New York: Harper & Row, 1970).
23. Jennifer Hansen, "The Art and Science of Reading Faces: Strategies of Racist Cinema in the Third Reich," *Shofar: An Interdisciplinary Journal of Jewish Studies* 28 (1, forthcoming).
24. Marvin Perry and Frederick M. Schweitzer, eds., *Jewish-Christian Encounters over the Centuries: Symbiosis, Prejudice, Holocaust, Dialogue* (New York: Peter Lang, 1994), p. 156.
25. Rohinton Mistry, *A Fine Balance* (New York: Vintage Books, 2001), p. 239.
26. Vikram Seth, *A Suitable Boy: A Novel* (New York: Harper Perennial, 2005), p. 217.
27. Gabriella Safran, *Rewriting the Jew: Assimilation Narratives in the Russian Empire* (Stanford, Calif.: Stanford University Press, 2000), p. 39.
28. Michael Wex, *Born to Kvetch: Yiddish Language and Culture in All of Its Moods* (New York: Macmillan, 2005), p. 179. On culturally manipulating tasty food to seem disgusting, see Carolyn Korsmeyer, "Delightful, Delicious, Disgusting," *Journal of Aesthetics and Art Criticism* 60 (3, Summer 2002): 217–25; Paul Rozin, "Development in the Food Domain," *Developmental Psychology* 26 (4, 1990): 555–62; Leann L. Birch, "Development of Food Preferences," *Annual Review of Nutrition* 19 (1999): 41–62.
29. In another context, these have also been called "imagined communities"; see Benedict Anderson, *Imagined Communities: Reflections on the Origins and Spread of Nationalism* (London: Verso, 1983). The connection between food and larger (early) social units is discussed in Susan Pollock, "Feasts, Funerals, and Fast Food in Early Mesopotamian States," in Tamara L. Bray, ed., *The Archaeology and Politics of Food and Feasting in Early States and Empires* (New York: Kluwer Academic, 2003), pp. 17–38.
30. Liz Wilson, "Beggars Can Be Choosers: Mahakassapa as a Selective Eater of Offering," in Jon Walters, Jacob Kinnard, and Ann Blackburn, eds., *Constituting Theravada Communities* (Albany: State University of New York Press, 2003); see also Liz Wilson, *Charming Cadavers: Horrific Figurations of the Feminine in Indian Buddhist Hagiographic Literature* (Chicago: University of Chicago Press, 1996).

31. Tertullian, *De Ieiunio*, 16:8, quoted in Veronica Grimm, *From Feasting to Fasting: The Evolution of a Sin* (London: Routledge, 1998), p. 135. Grimm remarks in general that, "at best, food and eating were tolerated as necessary for life . . . at worst, they were seen as the devil's snare" (p. 91). The theme of eating and sexual sin was familiar and explicit in the insistence on fasting; see Peter Brown, *The Body and Society: Men, Women, and Sexual Renunciation in Early Christianity* (New York: Columbia University Press, 1988).
32. Clement of Alexandria, *Christ the Educator (Paidagogos)*, trans. Simon P. Wood (New York: Fathers of the Church, 1954), 2:1:3–4.
33. Clement of Alexandria, *Christ the Educator*, 2:1:3–4.
34. Jerome, letter 54.10, quoted in Grimm, *From Feasting to Fasting*, p. 167. Such comments are very common in Jerome's work; see, for example, letter 22, reprinted in *The Letters of Saint Jerome*, vol. 1, trans. Charles Christopher Mierow (Westminster, Md.: Newman Press, 1963), pp. 137–43.
35. Saint Augustine, *Confessions*, 10.33, in *The Confessions of Saint Augustine*, trans. Henry Chadwick (New York: Oxford University Press, 1991).
36. Saint Augustine, *Confessions*, 10.31.43.
37. Saint Augustine, *Confessions*, 10.31.43.
38. For Philo on overeating, see Grimm, *From Feasting to Fasting*, p. 30.
39. Andrew Olendzki, trans., "King Pasendi Goes on a Diet," in Samyutta Nikaya ("Donapaka Sutta") 3.13, available at: http://www.accesstoinsight.org/tipitaka/sn/sn03/sn03.013.olen.html.
40. Patimokkha, Pakittiya Dhamma, 39, in T. W. Rhys Davids, trans., *Buddhist-sutras* (Delhi: Motilal Banarsidas, 1980), p. 40.
41. Patimokkha, 35–36, in Rhys Davids, trans., *Buddhist-sutras*, pp. 39–40.
42. *Milindapanha: The Questions of King Milinda*, trans. T. W. Rhys Davids (Delhi: Motilal Banarsidass, 1982), chs. 12, 17. Hindu negative food hedonics also began to emerge at the time of the rise of Buddhism—along with the rise of Hindu monastic (renunciatory) groups, as indicated in the Upanishads and similar sources.
43. The subject of social organization and the control of desire is extremely familiar in sociological literature. However, the area of control usually discussed is sexual behavior. See, for example, Victor Turner, *The Ritual Process: Structure and Anti-Structure* (Ithaca, N.Y.: Cornell University Press, 1969), p. 104.
44. The distinction is often put in terms of the society based on agape love versus the society based on conflict, competition, or what René Girard calls "mimetic desire"; see René Girard, *Violence and the Sacred* (Baltimore: Johns Hopkins University Press, 1979). For applications to the early Christian break from Judaism, see Robert G. Hamerton-Kelly, *Sacred Violence: Paul's Hermeneutic of the Cross* (Minneapolis: Fortress Press, 1992), p. 69.

45. On Perpetua and Felicitas, see chapter 9.
46. The concept of a moral group (or community) originated, I believe, with the French sociologist Émile Durkheim, who used it in his definition of religion (in *The Elementary Forms of the Religious Life*). My own discussion of social groups is loosely based on the pillars of the sociology of groups, particularly with help from the following: Ferdinand Tonnies, *Community and Association* (London: Routledge and Paul, 1955); Gunther Luschen and Gregory Stone, eds., *Herman Schmalenbach on Society and Experience* (Chicago: Chicago University Press, 1977); Talcott Parsons, *The Evolution of Societies* (Englewood Cliffs, N.J.: Prentice Hall, 1977); and Edward Shils, *Periphery and Center: Essays on Macrosociology* (Chicago: Chicago University Press, 1975).
47. The concept of the moral group or moral community is most often used to describe such organizations as churches, voluntary associations, charities, and so forth. This tends to be an evaluative and ethical concept; see, for example, Larry L. Rasmussen, *Moral Fragments and Moral Community: A Proposal for Church in Society* (Minneapolis: Fortress Press, 1993). But the term is also used in a sociological sense, as I use it, to describe groups that organize and coalesce around ideas more than organic principles or social structure; see, for example, Monica L. Smith, ed., *The Social Construction of Ancient Cities* (Washington, D.C.: Smithsonian Institution Press, 2003). Victor Turner speaks of liminal groups, or communitas, which often mirror normal social structure but are inverted; see *Ritual Structure*, chs. 3 and 4.
48. See Norbert Elias, *The Civilizing Process* (Oxford: Basil Blackwell, 1994); P. Ricoeur, "Guilt, Ethics, and Religion," in C. E. Ellis Nelson, ed., *Conscience: Theological and Psychological Perspectives* (New York: Newman, 1973); John Carroll, *Guilt: The Grey Eminence Behind Character, History, and Culture* (London: Routledge and Kegan Paul, 1985).
49. Ariel Glucklich, *Sacred Pain: Hurting the Body for the Sake of the Soul* (New York: Oxford University Press, 2001), especially ch. 7.
50. Émile Durkheim, in *The Elementary Forms of the Religious Life* (New York: Free Press, 1965), initiated the study of such rituals, which he called "piacular"—related to atonement or expiation. The role of pain in such rituals is discussed in detail in Glucklich, *Sacred Pain*.

Chapter 6: God's Love and the Prozac Effect

1. Statisticians would dearly love to understand how crowds predict—there is a lot of money to be made in understanding how stock money traders predict elections, market behavior, and, yes, even the weather in Florida (better than the National Weather Service); see James Surowiecki, *The Wisdom of Crowds: Why the Many Are Smarter Than the Few and How Collective Wisdom Shapes Business, Economies, Societies, and Nations* (New York:

Little, Brown, 2004); Gary Stix, "When Markets Beat the Polls," *Scientific American* (March 2008): 38–45.

2. On Drew Westen's research at Emory, see Michael Shermer, "The Political Brain," *Scientific American* (July 2006): 36. The hypothesis about unconscious confirmation bias is widely shared. See, for example, Shankara Vedantam's column on the subject in the *Washington Post*, January 30, 2006; and David P. Redlawsk, ed., *Feeling Politics: Emotion in Political Information Processing* (New York: Palgrave Macmillan, 2006). For a more general approach to the interface between emotion and thinking, see Lisa Feldman Barrett and Peter Salovey, eds., *The Wisdom in Feeling: Psychological Processes in Emotional Intelligence* (New York: Guilford Press, 2002). For a more technical discussion of hedonic influences on rationality, see Jonathan Baron, "Nonconsequentialist Decisions," *Behavioral Brain Science* 17 (1994): 1–42; and a response by Michel Cabanac, "The Evolutionary Point of View: Rationality Is Elsewhere: Commentary on J. Baron's 'Nonconsequentialist Decisions,'" *Behavioral Brain Research* 19 (1996): 322.
3. On the Kaddish and its consoling role, see Anita Diamant, *Saying Kaddish: How to Comfort the Dying, Bury the Dead, and Mourn as a Jew* (New York: Schocken Press, 1999).
4. The "new atheism" is a term associated mainly with the following works: Sam Harris, *The End of Faith: Religion, Terror, and the Future of Reason* (New York: W. W. Norton, 2004); Richard Dawkins, *The God Delusion* (New York: Houghton Mifflin, 2006); and Christopher Hitchens, *God Is Not Great: How Religion Poisons Everything* (New York: Hachette Book Group, 2007). The most measured and sophisticated response I have seen is John F. Haught, *God and the New Atheism: A Critical Response to Dawkins, Harris, and Hitchens* (Louisville: Westminster John Knox Press, 2008).
5. Paul Tillich, *The Courage to Be* (New Haven, Conn.: Yale University Press, 1952), p. 24.
6. The three adaptive types of pleasure, as discussed in chapter 2, are replenishment or novelty, mastery, and exploration. These are taken up later in this chapter.
7. Peter Kramer, *Listening to Prozac* (New York: Penguin, 1997).
8. On the difference between emotion and feeling, see Antonio Damasio, *The Feeling of What Happens: Body and Emotion in the Making of Consciousness* (San Diego: Harcourt Brace, 1999), ch. 2. See also Joseph LeDoux, *Synaptic Self: How Our Brains Become Who We Are* (New York: Penguin, 2002), ch. 8.
9. Bernard Weiner, *An Attributional Theory of Motivation and Emotion* (New York: Springer-Verlag, 1986); Thomas Shelley Duval, Paul Silvia, and Neal Lalwani, *Self-Awareness and Causal Attribution: A Dual Systems Theory* (Boston: Kluwer Academic Publishers, 2001). See also Alan M. Leslie, "A

Theory of Agency," in Dan Sperber, David Premack, and Ann James Premack, eds., *Causal Cognition: A Multidisciplinary Debate* (Oxford: Clarendon Press, 1995), pp. 121–41. Pascal Boyer is famous for assigning causal attribution a central role in the evolution of religion, but I believe it is safer to limit it to more specialized functions such as accounting for the cause of a specific feeling. For an excellent recent review of attribution theories, see Bertram F. Malle, *How the Mind Explains Behavior: Folk Explanations, Meaning, and Social Interaction* (Cambridge, Mass.: MIT Press, 2004).

10. Weiner, *An Attributional Theory of Motivation and Emotion*, ch. 5; Duval, Silvia, and Lalwani, *Self-Awareness and Causal Attribution*, pp. 51–52.
11. Duval et al., *Self-Awareness and Causal Attribution*, p. 8.
12. Prozac does not actually cause a feeling of euphoria, but I decided to use it for the attribution error after considering, then rejecting, the term "espresso effect."
13. Evelyn Underhill, *Mysticism* (London: Methuen, 1911); Underhill, *Practical Mysticism: A Little Book for Normal People* (New York: Vintage Books, 2003).
14. Underhill, *Practical Mysticism*, pp. xiv–xxv.
15. Saint John of the Cross, *The Ascent of Mount Carmel*, in Kieran Kavanaugh, ed., *Saint John of the Cross: Selected Writings* (New York: Paulist Press, 1987).
16. Saint John of the Cross, *The Ascent of Mount Carmel*, in Kavanaugh, ed., *Saint John of the Cross*, 1.1.4, p. 62.
17. Saint John of the Cross, *The Ascent of Mount Carmel*, in Kavanaugh, ed., *Saint John of the Cross*, 1.5.5, p. 70.
18. Saint John of the Cross, *The Ascent of Mount Carmel*, in Kavanaugh, ed., *Saint John of the Cross*, 1.5.7, p. 71.
19. Saint John of the Cross, *The Ascent of Mount Carmel*, in Kavanaugh, ed., *Saint John of the Cross*, 2.7.5, p. 95; 2.7.7, p. 96.
20. Saint John of the Cross, *The Ascent of Mount Carmel*, in Kavanaugh, ed., *Saint John of the Cross*, 2.26.7, p. 139.
21. Teresa of Ávila, *The Interior Castle*, ed. Kieran Kavanaugh and Otilio Rodriguez (New York: Paulist Press, 1979), p. 143.
22. Teresa of Ávila, *The Interior Castle*, ed. Kieran Kavanaugh and Otilio Rodriguez, p. 187.
23. Henry Suso, "The Life of the Servant," in Frank Tobin, trans., *The Exemplar: With Two German Sermons* (New York: Paulist Press, 1989), ch. 14, p. 87.
24. "Little Book of Eternal Wisdom," in Tobin, trans., *The Exemplar*, pp. 295–96.
25. "Little Book of Letters," letter 3, pp. 340–41.
26. Suso, "The Life of the Servant," in Tobin, trans., *The Exemplar*, chs. 31 and 32, pp. 126–31.
27. Suso, "The Life of the Servant," in Tobin, trans., *The Exemplar*, chs. 31 and 32, pp. 126–31.

28. Annemarie Schimmel, *Mystical Dimensions of Islam* (Chapel Hill: University of North Carolina Press, 1975), p. 100. For another complete Sufi version of the path with its stations, see Abu'l Qasim al-Qushayri, *The Sufi Book of Spiritual Ascent*, trans. Rabia Harris (Chicago: ABC International Group, 1997).
29. Margaret Smith, *Muslim Women Mystics: The Life and Work of Rabi'a and Other Women Mystics in Islam* (Oxford: One World, 1994), pp. 2–15.
30. Smith, *Muslim Women Mystics*, p. 78. Compare this to Saint John of the Cross exclaiming, in reference to that first step, "Ah, the sheer grace!"
31. Smith, *Muslim Women Mystics*, p. 102.
32. Smith, *Muslim Women Mystics*, p. 107.
33. Smith, *Muslim Women Mystics*, p. 126.
34. Smith, *Muslim Women Mystics*, p. 127.
35. Meister Eckhart, "Selected Sermons," in Karen J. Campbell, ed., *German Mystical Writings* (New York: Continuum, 1991), p. 138.
36. Most psychological and philosophical studies of mysticism focus on consciousness rather than attention; see, for example, Robert K. C. Norman, ed., *The Innate Capacity: Mysticism, Psychology, and Philosophy* (New York: Oxford University Press, 1998). This book is attacking the more influential view of mystical consciousness, which can be found in books such as Steven T. Katz, ed., *Mysticism and Religious Traditions* (New York: Oxford University Press, 1983).
37. Paula M. Niedenthal and Shiobu Kitaya, eds., *The Heart's Eye: Emotional Influences in Perception and Attention* (San Diego: Academic Press, 1994), p. 8.
38. Harold E. Pashler, *The Psychology of Attention* (Cambridge, Mass.: MIT Press, 1998); Trevor W. Robbins, "Arousal and Attention: Psychopharmacological and Neuropsychological Studies in Experimental Animals," in Raja Parasuram, ed., *The Attentive Brain* (Cambridge, Mass.: MIT Press, 1999), pp. 189–220.
39. Edward B. Titchner, *Lectures on the Elementary Psychology of Feeling and Attention* (New York: Macmillan, 1908); William James, *The Principles of Psychology*, vol. 1 (New York: Dover, 1890).
40. See, for example, Donald E. Broadbent, *Perception and Communication* (London: Pergamon, 1958); D. A. Norman, "Toward a Theory of Memory and Attention," *Psychological Review* 75 (1968): 522–36.
41. Victor S. Johnston, "The Origin and Function of Pleasure," in James A. Russell, ed., *Pleasure* (East Sussex, U.K.: Psychology Press, 2003), pp. 167–80; see also Charles S. Carver, "Pleasure as a Sign You Can Attend to Something Else: Placing Positive Feelings Within a General Model of Effect," in Russell, *Pleasure*, pp. 241–62. A wonderful recent book on the joys of making decisions is Jonah Lehrer's *How We Decide* (New York: Houghton Mifflin Harcourt, 2009).

42. Douglas Derryberry and Don M. Tucker, "Motivating the Focus of Attention," in Niedenthal and Kitaya, *The Heart's Eye*, pp. 167–96.
43. On priming, see Pashler, *The Psychology of Attention*, pp. 67–69; Shinobu Kitayama and Susan Howard, "Affective Regulation of Perception and Comprehension: Amplification and Semantic Priming," in Niedenthal and Kitayama, *The Heart's Eye*, pp. 41–67.
44. This work was done at the Beckman Visual Cognition Lab at the University of Illinois and can be seen at: http://viscog.beckman.uiuc.edu/grafs/demos/15.html (opens in a new window).
45. Andrew B. Newberg and Stephanie K. Newberg, "The Neurobiology of Religious and Spiritual Experience," in Raymond F. Paloutzian and Crystal L. Park, eds., *Handbook of the Psychology of Religion and Spirituality* (New York: Guilford Press, 2005), pp. 199–215; see also Andrew Newberg, Eugene D'Aquili, and Vince Rause, *Why God Won't Go Away* (New York: Ballantine, 2002).
46. The human "self" is one of the most complex products of brain function and is far from fully understood. For a recent theory, see LeDoux, *Synaptic Self*, ch. 11. I have shown in a previous book (*Sacred Pain*) how mystics utilize massive overstimulation, in the form of pain, to diminish the sense of self.
47. R. D. Ranade, *Mysticism in India: The Poet-Saints of Maharashtra* (Albany: State University of New York Press, 1983), p. 125.
48. Anthony J. Steinbock, *Phenomenology and Mysticism: The Verticality of Religious Experience* (Bloomington: Indiana University Press, 2007). Steinbock follows in the footsteps of theologians and philosophers who apply the concept of verticality to religious experience (see pp. 11–12). I limit myself to brain states.
49. Saint Augustine, *Confessions*, 7.10.16, quoted in Bernard McGinn, *The Foundations of Mysticism: Origins to the Fifth Century* (New York: Crossroad, 1995), p. 233.
50. Ellen Berscheid, "Searching for the Meaning of Love," in Robert J. Sternberg and Karin Weis, eds., *The New Psychology of Love* (New Haven, Conn.: Yale University Press, 2006), pp. 171–83; see also Joanne Brown, *A Psychosocial Exploration of Love and Intimacy* (New York: Palgrave Macmillan, 2006).
51. Clyde Hendrick and Susan S. Hendrick, "Styles of Romantic Love," in Sternberg and Weis, *The New Psychology*, p. 153; see also Stephen G. Post, Lynn G. Underwood, Jeffrey P. Schloss, and William B. Hurlbut, eds., *Altruism and Altruistic Love: Science, Philosophy, and Religion in Dialogue* (New York: Oxford University Press, 2002); see especially the essays in part 2 by Lynn Underwood and in part 3 by Stephen Pope. See also Mark

Patrick Hederman, *Love Impatient, Love Unkind: Eros Human and Divine* (New York: Crossroad Publications, 2004).

52. Smith, *Muslim Women Mystics*, p. 79.
53. Helen Fisher, "The Drive to Love," in Sternberg and Weis, *The New Psychology*, p. 90; see also Fisher, *Why We Love: The Nature and Chemistry of Romantic Love* (New York: Henry Holt, 2004).
54. Kent C. Berridge, "Pleasure, Pain, Desire, and Dread: Hidden Core Processes of Emotion," in Daniel Kahneman, Ed Diener, and Norbert Schwarz, eds., *Well-being: The Foundations of Hedonic Psychology* (New York: Russell Sage Foundation, 1999), pp. 525–58. For more on this, see chapter 8 on heaven.
55. Fisher, "The Drive to Love," p. 90.
56. Hendrick and Hendrick, "Styles of Romantic Love," p. 153; Stephen G. Post, "The Tradition of Agape," in Post et al., *Altruism and Altruistic Love*, pp. 51–64.
57. Nadia Bolz-Weber, *The Sarcastic Lutheran*, quoted in Becky Garrison, *The New Atheist Crusaders and Their Unholy Grail: The Misguided Quest to Destroy Your Faith* (Nashville: Thomas Nelson, 2007), p. 172.

Chapter 7: Spiritual Love and the Seeds of Annihilation

1. Martin Buber, *I and Thou* (New York: Simon & Schuster, 1970).
2. Rashi Chumash commentary on Genesis 22:12, in A. M. Silberman and M. Rosenbaum, *Chumash with Rashi's Commentary* (Nanuet, N.Y.: Phillip Feldheim Inc., 1985).
3. Lacan discusses this episode in "Introduction to The Names of the Father Seminar," *October* 40 (Spring 1987): 81–95.
4. See Eyal Dotan, "Will This Satisfy You, God?" (book review of a new translation of Lacan's seminar into Hebrew), *Haaretz*, Sefarim, May 24, 2006, p. 10.
5. Max Weber, *On Charisma and Institution Building: Selected Papers* (Chicago: University of Chicago Press, 1968).
6. See, for example, the 2007 flyer at: http://swamahiman.org/assets/Houston.pdf.
7. Quoted in Gershom Scholem, "The Historical Figure of R. Israel Baal-Shem-Tov," in David Assaf, ed., *Zaddik and Devotees: Historical and Sociological Aspects of Hasidism* (Jerusalem: Zalman Shazar, 2001), p. 82. See also Joseph Dan, *The Hasidic Story: Its History and Development* (Jerusalem: Keter, 1990).
8. On the Jewish mystic R. Yehiel Mikhel and his love-based group, for example, see Mor Altshuler, *The Messianic Secret of Hasidism* (Leiden: Brill, 2006), pp. 108–11.

9. Franklin D. Lewis, *Rumi: Past and Present, East and West* (Oxford: One World, 2008).
10. Quoted in Lewis, *Rumi*, p. 153.
11. Quoted in Lewis, *Rumi*, p. 153.
12. Quoted in Lewis, *Rumi*, p. 168.
13. Quoted in Lewis, *Rumi*, p. 169.
14. Josef W. Meri, *The Cult of Saints Among Muslims and Jews in Medieval Syria* (Oxford: Oxford University Press, 2002), p. 88. See also Lewis, *Rumi*, pp. 26, 410.
15. Quoted in Norman Lamm, *The Religious Thought of Hasidism: Text and Commentary* (Hoboken, N.J.: Yeshiva University Press, 1999), p. 299; see also p. 298.
16. William K. Mahony, "The Guru-Disciple Relationship: The Context for Transformation," in Douglas Renfrew Brooks et al., eds., *Meditation Revolution: A History and Theology of the Siddha Yoga Lineage* (South Fallsburg, N.Y.: Agama Press, 1997), pp. 260–61.
17. Edward C. Dimock, *Caitanya Caritamrta of Krsnadasa Kaviraja: A Translation and Commentary* (Cambridge, Mass.: Harvard University Press, 1999).
18. For a tiny sample, see Meri, *The Cult of Saints*, p. 79; Lise McKean, *Divine Enterprise: Gurus and the Hindu Nationalist Movement* (Chicago: University of Chicago Press, 1996), pp. 22–23; Paul E. Muller-Ortega, "The Siddhas: Paradoxical Exemplar of Indian Spirituality," in Brooks et al., *Meditation Revolution*, p. 189; Lamm, *The Religious Thought of Hasidim*, p. 255; Dan, *The Hasidic Story*, p. 103.
19. W. Y. Evans-Wentz, ed., *Tibet's Great Yogi Milarepa: A Biography from the Tibetan*, 2nd ed. (London: Oxford University Press, 1973); Lobsang P. Lhalungpa, trans., *The Life of Milarepa* (Boulder, Colo.: Shambhala, 1977).
20. Victor Turner, *The Ritual Process: Structure and Anti-Structure* (Ithaca, N.Y.: Cornell University Press, 1977).
21. Abdellah Hammoudi, *Master and Disciple: The Cultural Foundations of Moroccan Authoritarianism* (Chicago: University of Chicago Press, 1997), pp. 87–92.
22. Carl W. Ernst and Bruce B. Lawrence quoting Annemarie Schimmel's conference paper "Sufi Biographies," in *Sufi Martyrs of Love: The Chishti Order in South Asia and Beyond* (New York: Palgrave Macmillan, 2002), p. 71.
23. For such techniques, see, for example, Carl W. Ernst and Bruce B. Lawrence, *Sufi Martyrs of Love: The Chishti Order in South Asia and Beyond* (New York: Palgrave Macmillan, 2002), p. 30.
24. Margaret Malamud, "Gender and Spiritual Self-Fashioning: The Master-Disciple Relationship in Classical Sufism," *Journal of the American Academy of Religion* 64 (1, 1996): 89–117. Malamud is citing Najm al Din al-Razi, *The Path of God's Bondsmen from Origin to Return.*

25. Mendel Feikazh, "Hasidism: A Socio-Religious Movement in Light of Devekut," in Assaf, *Zaddik and Devotees*, p. 445.
26. Quoted in Lamm, *The Religious Thought of Hasidism*, p. 205.
27. Robert Svoboda, *Aghora: At the Left Hand of God* (Albuquerque, N.M.: Brotherhood of Life, 1986), p. 41, quoted in McKean, *Divine Enterprise*, p. 3. The tropes of gender (passive = feminine) are central not only to the guru-disciple relationship but also to the conception of the conflict between the religious hero and his enemy. This indirect relation between love and violence requires further study; see Sikata Banerjee, *Make Me a Man!: Masculinity, Hinduism, and Nationalism in India* (Albany: State University of New York Press, 2005).
28. According to Zeev Gris, an "erotic myth," which expressed the intimate friendship of the few, was used to fashion a social ethos for the larger group of the Hasidim who gather around the Tzadik; see "From Myth to Ethos," 134–35, cited in Haviva Pedaya, "The Development of the Social-Religious-Economic Model in Hasidism: The Pidyon, the Havura, and the Pilgrimage," in Assaf, *Zaddik and Devotees*, p. 359.
29. Quoted in Lamm, *The Religious Thought of Hasidism*, p. 301.
30. Lewis, *Rumi*, p. 28; Meri, *The Cult of Saints*, p. 91.
31. Swami Muktananda, quoted in Douglas Renfrew Brooks, "The Canons of Siddha Yoga: The Body of Scripture and the Form of the Guru," in Douglas Renfrew Brooks et al., eds., *Meditation Revolution: A History and Theology of the Siddha Yoga Lineage* (South Fallsburg, N.Y.: Agama Press, 1997), p. 332.
32. Leonard W. Levy, *Blasphemy: Verbal Offense Against the Sacred from Moses to Salman Rushdie* (Chapel Hill: University of North Carolina Press, 1993), p. 171.
33. Havivah Pedaya, "The Development of the Social-Religious-Economic Model in Hasidism: The Pidyon, the Havura, and the Pilgrimage," in Assaf, *Zaddik and Devotees*, p. 344, n. 4.
34. Weber, *On Charisma*, p. 22. See also Clifford Geertz, *Local Knowledge* (New York: Basic Books, 1983), pp. 125–34; Pnina Werbner, *Pilgrims of Love: The Anthropology of a Global Sufi Cult* (Bloomington: Indiana University Press, 2003), p. 25.
35. Gerhard Gershom Scholem, *Sabbatai Sevi: The Mystical Messiah, 1626–1676* (Princeton, N.J.: Princeton University Press, 1973); Yehuda Liebes, *On Sabbateanisn and Its Kabbalah: Collected Essays* (Jerusalem: Bialik Institute, 1995).
36. Richard Lawrence Hock, "The Politics of Redemption: Rabbi Tzvi Yehudah ha-Kohen Kook and the Origins of Gush Emunim" (PhD diss., University of California at Santa Barbara, 1994), ch. 4.
37. Peter van der Veer, *Religious Nationalism: Hindus and Muslims in India* (Berkeley: University of California Press, 1994), p. 34.

38. Arthur F. Buehler, *Sufi Heirs of the Prophet: The Indian Naqshbandiyya and the Rise of the Mediating Sufi Shaykh* (Columbia: University of South Carolina Press, 1998), p. 217.
39. Buehler, *Sufi Heirs of the Prophet*, p. 199.
40. Buehler, *Sufi Heirs of the Prophet*, pp. 8–9.
41. Ahmet Yasar Ocak, "Sufi Milieux and Political Authority in Turkish History: A General Overview (Thirteenth–Seventeenth Centuries)," in Paul L. Heck, ed., *Sufism and Politics* (Princeton, N.J.: Marcus Weiner, 2007), p. 186.
42. Hammoudi, *Master and Disciple*.
43. Lydia Khalil, "Al-Qaeda and the Muslim Brotherhood: United by Strategy, Divided by Tactics," *Terrorism Monitor* 4 (6, 2006): 7–9.
44. Heck, *Sufism and Politics*; van der Veer, *Religious Nationalism*; Werbner, *Pilgrims of Love*.
45. Fadhil Ali, "Sufi Insurgent Groups in Iraq," *Terrorism Monitor* 6 (2, 2008), available at: http://www.jamestown.org. This journal and others such as *Terrorism Focus* and *Spotlight on Terror* are excellent sources of material on the shifting landscape of religious terrorism. Some of the groups listed include: Jaysh Rijal al-Tariqa al-Naqshbandi (JRTN), the largest Sufi insurgent group; Kitabat al-Sheikh abd-al-Qudir al-Jilanin al-Jihadia (the Jihadi Battalion of Sheikh Abd al-Qadir al-Jilani); and the Sufi Squadron of Sheikh Abd al-Qadir at Mosul. See also Fadhil Ali, "The Ansar al-Mahdi and the Continuing Threat of the Doomsday Cults in Iraq," *Terrorism Monitor* 6 (4, 2008): 7–9. The leaders of the non-Sufi but messianic groups discussed include: Ahamad al-Hassan, aka Shaykh al-Yamani, head of Ansar al-Mahdi (Helpers of the Expected One), a Shi'te, non-Sufi, but apocalyptic militant group that operates in southern Iraq; Dhia Abdul Zahra al-Gar'awi, who leads Jund al-Samaa (Soldiers of Heaven), also operating in southern Iraq; and Moqtada al-Sadr, who heads the Jaysh al-Mahdi, is the best-known leader of such Shi'ite groups, and gives every appearance of being nonimmolative.
46. See Lilly Weissbrod, "Coping with the Failure of a Prophecy: The Israeli Disengagement from the Gaza Strip," *Journal of Religion and Society* 10 (2008), available at: http://moses.creighton.edu/JRS/toc/2008.html; see also the discussion in chapter 3 of cognitive dissonance and Leon Festinger.
47. McKean, *Divine Enterprise*, p. 31; see also van der Veer, *Religious Nationalism*, p. 46; Christophe Jaffrelot, ed., *Hindu Nationalism: A Reader* (Princeton, N.J.: Princeton University Press, 2007).
48. Isabel Kershner, "Radical Settlers Take on Israel," *New York Times*, September 26, 2008.
49. Ross Moret, "Potential for Apocalypse: Violence and Eschatology in the Israeli-Palestinian Conflict," *Journal of Religion and Society* 10 (2008), avail-

able at: http://moses.creighton.edu/JRS/toc/2008.html. I have discussed Juergensmeyer elsewhere in this book, but see Thomas Robbins, "Sources of Volatility in Religious Movements," in David G. Bromley and Gordon J. Melton, eds., *Cults, Religion, and Violence* (Cambridge: Cambridge University Press, 2002), pp. 57–79.

Chapter 8: The Paradise Paradox: The Misery of Heaven-Addicts

1. Quoted (from the Robert the Monk version) in James Harvey Robinson, ed., *Readings in European History*, vol. 1 (Boston: Ginn and Co., 1904), pp. 312–16.
2. Osama bin Laden's "Declaration of War Against the Americans" was first published in *Al Quds Al Arabi*, August 1996.
3. Other measurements have been roughly similar. A 1997 Opinion Dynamics Survey for *Fox News* found that a whopping 88 percent of all Americans believe that heaven is a real place.
4. J. Krishnamurti, *Total Freedom: The Essential Krishnamurti* (San Francisco: HarperSanFrancisco, 1996), p. 314.
5. Krishnamurti, *Total Freedom*, p. 314.
6. For details on what Krishnamurti means by self and how it is constructed in experience, see J. Krishnamurti, *The First and Last Freedom* (Wheaton, Ill.: Quest, 1971), ch. 9. Buddha taught the Four Noble Truths, the second being that the cause of suffering is craving (*tanha*—the desire for objects and for becoming); see Dhammacakkapavattana SN 56.11.
7. Paul Tillich, "Existential Analyses and Religious Symbols," in Harold A. Basilius, ed., *Contemporary Problems in Religion* (Detroit: Wayne University Press, 1956), pp. 53–54.
8. Few have made this argument better than Friedrich Nietzsche, for example, in *Beyond Good and Evil: Prelude to a Philosophy of the Future.* Just as entertaining for a contemporary sensibility is Christopher Hitchens, *God Is Not Great: How Religion Poisons Everything* (New York: Twelve, 2007), ch. 3.
9. Colleen McDannell and Bernhard Lang, *Heaven: A History* (New Haven, Conn.: Yale University Press, 1988); Jeffrey Burton Russell, *A History of Heaven: The Singing Silence* (Princeton, N.J.: Princeton University Press, 1997); Regis Martin, *The Last Things: Death, Judgment, Hell, Heaven* (San Francisco: Ignatius Press, 1998); Ra'anan S. Boustan and Annette Yoshiko Reed, *Heavenly Realms and Earthly Realities in Late Antique Religions* (New York: Cambridge University Press, 2004).
10. Saint Augustine, *Confessions*, 9.10.24, in *The Confessions of Saint Augustine*, trans. Henry Chadwick (New York: Oxford University Press, 1991).
11. McDannell and Lang, *Heaven*, pp. 16–17.

12. Otfried of Weissenburg, a ninth-century German monk and poet, quoted in McDannell and Lang, *Heaven*, p. 70; see also Jean Delumeau, *History of Paradise: The Garden of Eden in Myth and Tradition* (Urbana: University of Illinois Press, 2000).
13. Dante Alighieri, "Paradise" 30.60, *The Divine Comedy*, trans. Henry F. Cary, Harvard Classics vol. 20 (New York: P. F. Collier & Son Co., 1909–14).
14. Both Hindus and Buddhists believe in karma and reincarnation as alternative forms of afterlife and often combine these with heaven—which would make heaven a temporary place where good merit keeps us until it is all used up. For Buddhists, see Akira Sadakata, *Buddhist Cosmology, Philosophy, and Origins* (Tokyo: Kosei, 1997); for Hindus, see David M. Knipe, "Hindu Eschatology," in Jerry Walls, ed., *The Oxford Handbook of Eschatology* (New York: Oxford University Press, 2008), pp. 170–90.
15. Mahabharata, Book of the Assembly Hall (2), 246.30–36.
16. McDannell and Lang, *Heaven*, p. 353.
17. For more on this joy, see chapter 7 on divine love.
18. The idea that surveillance makes us miserable, that it is connected to the development of modern guilt, is very familiar to readers of the French theorist Michel Foucault, primarily through his work *Discipline and Punish: The Birth of the Prison* (New York: Vintage, 1995). I too have contributed to this cottage industry, in *Sacred Pain*, ch. 7 (on the Inquisition).
19. Carolyn R. Morillo, *Contingent Creatures: A Reward Event Theory of Motivation and Value* (Lanham, Md.: Littlefield Adams Books, 1995), p. 130. In technical terms, the calculation behind pleasure must take into account expected utility, which depends on already knowing what the pleasure is; see Paul W. Glimcher, *Decisions, Uncertainty, and the Brain* (Cambridge, Mass.: MIT Press, 2003), pp. 325–26.
20. According to Max Weber (*The Protestant Ethic and the Spirit of Capitalism*) and the legions of commentators on his work, American capitalism is something like such a solution.
21. K. W. Spence, *Behavior Theory and Conditioning* (New Haven, Conn.: Yale University Press, 1956). Like earlier researchers, Spence found that reinforcement works only with simple tasks and is actually detrimental to complex ones.
22. Barry Schwartz, *The Paradox of Choice: Why More Is Less* (New York: Harper Perennial, 2005). This does not mean that punishment works any better, of course. For one example in a vast literature, see Karl Sigmund, Christopher Hauert, and Martin A. Nowak, "Reward and Punishment," *Proceedings of the National Academy of Science of the United States of America* 96 (19, 2002): 10757–62. According to the study, when punishment or re-

wards is measured, prosocial behavior reputation appears to be the most important factor.

23. Mark R. Lepper and David Greene, eds., *The Hidden Costs of Reward: New Perspectives on the Psychology of Human Motivation* (Hillsdale, N.J.: Lawrence Erlbaum Associates, 1978).
24. Edward L. Deci, *Why We Do What We Do: Understanding Self-Motivation* (New York: Penguin, 1996). Much of this research began to be published in the 1970s. See, for example, M. Ross, R. Karniol, and M. Rothstein, "Reward Contingency and Intrinsic Motivation in Children: A Test of the Delay of Gratification Hypothesis," *Journal of Personality and Social Psychology* 33 (1976): 442–47; Mark R. Lepper and David Greene, "Overjustification Research and Beyond: Toward a Means-Ends Analysis of Intrinsic and Extrinsic Motivation," in Lepper and Greene, *The Hidden Costs*, pp. 109–48. For more recent work, see Bridget S. O'Brien and Paul J. Frick, "Reward Dominance: Associations with Anxiety, Conduct Problems, and Psychopathy in Children," *Journal of Abnormal Child Psychology* 24 (2, 1996): 223–40; Gregory-Adrian Offringa, "Effects of Reward Distribution in Task Groups on Member Performance, Self-esteem, and State Anxiety," *Dissertation Abstracts International* 57 (10-b, 1997): 6586. For a recent survey, see Moshe Zeidner and Gerald Matthews, "Evaluation Anxiety: Current Theory and Research," in Andrew J. Elliot and Carol S. Dweck, eds., *Handbook of Competence and Motivation* (New York: Guilford Press, 2005), pp. 141–66.
25. Alfie Kohn, *Punished by Rewards: The Trouble with Gold Stars, Incentive Plans, A's, Praise, and Other Bribes* (Boston: Houghton Mifflin, 1993).
26. Arthur Green, trans., *Menachem Nahum of Chernobyl: Upright Practices, The Light of the Eyes* (Mahwah, N.J.: Paulist Press, 1982), pp. 249–50.
27. See "Are We Happy Yet?" the February 13, 2006, Pew survey on the correlation between happiness and church attendance, available at: http://pewresearch.org/pubs/301/are-we-happy-yet.
28. Kent C. Berridge, "Pleasures of the Brain," *Brain and Cognition* 52 (2003): 106–28.
29. Jon Elster, ed., *Addiction: Entries and Exits* (New York: Russell Sage Foundation, 1999).
30. Vincent J. Miller, *Consuming Religion: Christian Faith and Practice in a Consumer Culture* (New York: Continuum, 2005).
31. Jon Elster, "Emotion and Addiction: Neurobiology, Culture, and Choice," in Elster, *Addiction*, pp. 239–76.
32. Glimcher, *Decisions, Uncertainty, and the Brain*, ch. 11.
33. On the meaning of social capital (honor, love, respect, and other prized values obtained in a social context) and the rationality of the choices made

by religious actors, see Rodney Stark and Roger Finke, *Acts of Faith: Explaining the Human Side of Religion* (Berkeley: University of California Press, 2000). Stark and his colleagues (William Bainbridge, Laurence Iannaccone, and others) have been systematically applying rational choice theory to religion over the last few decades. A detailed application of this method to a single case can be seen in Susan Kwilecki and Loretta S. Wilson, "Was Mother Teresa Maximizing Her Utility? An Idiographic Application of Rational Choice Theory," *Journal for the Scientific Study of Religion* 37 (2, June 1998): 205–21. This subject is discussed in greater detail in connection with obedience and love in chapter 8; see Marcel Mauss, *The Gift: The Form and Reason for Exchange in Archaic Societies* (New York: W. W. Norton, 1990).

Chapter 9: The Martyr's Theater

1. The topic of martyrdom has inspired a vast amount of secondary literature. A few of the works I have consulted for the definition include: Samuel Z. Klausner, "Martyrdom," in Mircea Eliade, ed., *Encyclopedia of Religion* (New York: Macmillan, 1987); G. W. Bowersock, *Martyrdom and Rome* (Cambridge: Cambridge University Press, 1995); Jan Willem Van Henten and Friedrich Avemarie, *Martyrdom and Noble Death: Selected Texts from Graeco-Roman, Jewish, and Christian Antiquity* (London: Routledge, 2002); Daniel Boyarin, *Dying for God: Martyrdom and the Making of Christianity and Judaism* (Palo Alto, Calif.: Stanford University Press, 1999); Susan Bergman, *Martyrs* (San Francisco: HarperSanFrancisco, 1996); W. H. C. Frend, *Martyrdom and Persecution in the Early Church: A Study of Conflict from the Maccabees to Donatus* (Oxford: Oxford University Press, 1965).
2. Shmuel Shepkaru, *Jewish Martyrs in the Pagan and Christian Worlds* (New York: Cambridge University Press, 2006). On Akiva and the contemporary martyr Hananiya ben Teradyon, see pp. 73–78.
3. According to Philippe Buc, Christian sects "hijacked" Roman civic rituals of criminal execution for their own purpose of promoting the condemned as martyrs; see *The Dangers of Ritual: Between Early Medieval Texts and Social Scientific Theory* (Princeton, N.J.: Princeton University Press, 2001), p. 124.
4. Boyarin, *Dying for God*, p. 116. In truth, martyrdom consists of several distinct but related genres, which are explored in Van Henten and Avemarie, *Martyrdom and Noble Death*, pp. 5–6. For a carefully analyzed literary construction of a Muslim martyr, see Carl W. Ernst, "From Hagiography to Martyrology: Conflicting Testimonies to a Sufi Martyr of the Delhi Sultanate" *History of Religions* 24 (4, May 1985): 308–27.
5. There will be more on theater later in the chapter where the discussion,

largely anthropological and sociological, draws on Victor Turner, *From Ritual to Theater* (New York: Performing Arts Journal Publications, 1982).

6. On the costumes worn by the executed, see K. M. Coleman, "Fatal Charades: Roman Executions Staged as Mythological Enactment," *Journal of Roman Studies* 50 (1990): 44–73.
7. *The Passion of SS. Perpetua and Felicity: Together with the Sermons of Saint Augustine on These Saints*, trans. W. H. Shewring (London: Sheed and Ward, 1931).
8. Rex D. Butler, *The New Prophecy and "New Visions": Evidence of Montanism in the Passion of Perpetua and Felicitas* (Washington, D.C.: Catholic University of America Press, 2006).
9. *The Passion of SS. Perpetua and Felicity*, p. 27.
10. *The Passion of SS. Perpetua and Felicity*, p. 27.
11. *The Passion of SS. Perpetua and Felicity*, p. 37.
12. Coleman shows in "Fatal Charades" (p. 59) how the scripting of the execution was distorted in the telling to match the martyr's prediction of his own death.
13. There have been many dissenting Christian voices against the celebration of martyrs. My favorite is that of Dag Hammarskjold, the mystical author and UN secretary-general who remarked that with martyrs the great commitment obscures the little one and that self-sacrifice can lead to a hardening of the heart; see *Markings* (New York: Knopf, 1964), p. 131. This observation seems particularly relevant to the case of Perpetua.
14. Lacy Baldwin Smith, *Fools, Martyrs, Traitors: The Story of Martyrdom in the Western World* (New York: Knopf, 1997), p. 102; A. D. Nock, *Conversion: The Old and the New in Religion from Alexander the Great to Augustus of Hippo* (Oxford: Oxford University Press, 1933), pp. 199–200.
15. Smith, *Fools, Martyrs, Traitors*, pp. 102–3.
16. Judith Butler, *Antigone's Claim: Kinship Between Life and Death* (New York: Columbia University Press, 2000), pp. 3–4.
17. Turner, *From Ritual to Theater*, p. 69.
18. Ariel Glucklich, *The End of Magic* (Oxford: Oxford University Press, 1997), pp. 49–53; Sudhir Kakar, *Shamans, Mystics, and Doctors: A Psychological Inquiry into India and Its Healing Traditions* (New York: Knopf, 1982).
19. Walter Burkert, *Greek Religion*, trans. John Raffan (Cambridge, Mass.: Harvard University Press, 1985). Not all scholars agree on the ritual-to-theater connection in Greece, and classicists are more likely to reject the idea than anthropologists and historians; see, for example, Kimberly C. Patton, "Discussion," in Eric Caputo and Margaret C. Miller, eds., *The Origins of Theater in Ancient Greece and Beyond: From Ritual to Drama* (New York: Cambridge University Press, 2007), p. 371.
20. Burkert, *Greek Religion*, pp. 190–93.
21. Burkert, *Greek Religion*, p. 106.

22. Carlin A. Barton, *The Sorrows of the Ancient Romans: The Gladiator and the Monster* (Princeton, N.J.: Princeton University Press, 1993).
23. René Girard, *Violence and the Sacred*, trans. Patrick Gregory (Baltimore: Johns Hopkins University Press, 1979).
24. Norbert Elias, *On Civilization, Power, and Knowledge: Selected Writings* (Chicago: University of Chicago Press, 1998), p. 171.
25. Carolyn Marvin and David W. Ingle, "Blood Sacrifice and the Nation: Revisiting Civil Religion," *Journal of the American Academy of Religion* 64 (4, Winter 1996): 767–780. An Orthodox Jew who once tried to convert me asked me, as a starter, what I believed in so deeply that I would die for it.
26. Barton, *The Sorrows of the Ancient Romans*, p. 14.
27. Philippe Buc (*The Dangers of Ritual*, p. 135) quotes Dionysius Bishop of Alexandria as exalting "martyrdom for the sake of preventing division [within the Church]" beyond dying for refusing to worship idols.
28. The death of the martyr in the ring is an agonistic victory over the devil; see Buc, *The Dangers of Ritual*, pp. 137–38.
29. Peter Chelkowski, *Ta'ziyeh: Ritual and Drama in Iran* (New York: New York University Press, 1979); Jamshid Malekpour, *The Islamic Drama* (London: Frank Cass Publishers, 2004).
30. Malekpour, *The Islamic Drama*, pp. 26–27.
31. Sadeq Humayuni, "An Analysis of the Ta'ziyeh of Qasem," in Chelkowski, *Ta'ziyeh*, p. 21.
32. Malekpour, *The Islamic Drama*, pp. 20–21. On the connection between drama, martyrdom, and political mobilization in Iran, see Manochehr Dorraj, "Symbolic and Utilitarian Political Value of a Tradition: Martyrdom in the Iranian Political Culture," *Review of Politics* 59 (3, 1997): 489–521.
33. Martyrdom as resistance or struggle (*jihad*) goes far beyond Shi'ite Islam and plays a central role in the Qur'an: "And those slain in the way of God, He will not send their works astray. He will guide them, and dispose their minds aright, and He will admit them to Paradise" (47:4–6). This is quoted in Keith Lewinstein, "The Revaluation of Martyrdom in Early Islam," in Margaret Cormack, ed., *Sacrificing the Self: Perspectives on Martyrdom and Religion* (New York: Oxford University Press, 2002), p. 80. For the effects of secularization on religious action in the West, see Jose Casanova, *Public Religions in the Modern World* (Chicago: University of Chicago Press, 1994).
34. Francisco R. Adrados, *Festival, Comedy, and Tragedy: The Greek Origins of Theater* (Leiden: E. J. Brill, 1975), pp. 11–12, 86; see also Max Harris, *Carnival and Other Christian Festivals: Folk Theology and Folk Performance* (Austin: University of Texas Press, 2003); Meg Twycross, ed., *Festive Drama: Papers from the Sixth Triennial Colloquium of the International Society*

for the Study of Medieval Theater (Cambridge: D. S. Brewer, 1996); Virgil Nemoianu and Robert Royal, eds., *Play, Literature, Religion: Essays in Cultural Intertextuality* (Albany: State University of New York Press, 1992). The ludic performance—not necessarily comic but absurdly theatrical—at times reduced violence against Jews in medieval Europe by performing it ritually as a sort of street play; see David Nirenberg, *Communities of Violence: Persecution of Minorities in the Middle Ages* (Princeton, N.J.: Princeton University Press, 1996), ch. 7.

35. K. J. Dover, *Aristophanic Comedy* (Berkeley: University of California Press, 1972), p. 67.
36. The horse sacrifice, to give one example, featured verbal jousting of a risqué nature between the priests and the queens; see Ariel Glucklich, *The Strides of Vishnu* (New York: Oxford University Press, 2008), ch. 7. A detailed study of comedy in South India can be seen in David Dean Shulman, *The King and the Clown in South Indian Myth and Poetry* (Princeton, N.J.: Princeton University Press, 1985).
37. Jan Walsh Hokenson, *The Idea of Comedy: History, Theory, Critique* (Madison, N.J.: Fairleigh Dickinson University Press, 2006).
38. Robert Torrance, *The Comic Hero* (Cambridge, Mass.: Harvard University Press, 1978), p. 157.
39. Henri Bergson, *Laughter*, trans. Cloudesley Brereton and Fred Rothwell (New York: Macmillan, 1928), p. 150. The gap between Don Quixote's delusions and the perfectly sensible way he goes about executing his goals is the source of the comic absurdity.
40. Susan Purdie, *Comedy: The Mastery of Discourse* (Toronto: University of Toronto Press, 1993). Christopher Butler quotes Arthur Koestler's definition of jokes as "a clash of two mutually incompatible codes, or associative contexts which explodes the tension of their narrative"; see *Pleasure and the Arts* (Oxford: Clarendon Press, 2004), p. 4.
41. George de Forest Lord, *Heroic Mockery: Variations on Epic Themes from Homer to Joyce* (Newark: University of Delaware Press, 1977), p. 21.
42. Shabda 3, in Linda Hess and Shukdev Singh, trans., *The Bijak of Kabir* (New York: Oxford University Press, 2002), p. 42.
43. Kabir Parachai 7.15, in David N. Lorenzon, trans., *Kabir Legends and Ananta Das's Kabir Parachai* (Albany: State University of New York Press, 1991).
44. Kabir Granthavali 16.1, in John Stratton Hawley and Mark Juergensmeyer, trans., *Songs of the Saints of India* (New York: Oxford University Press, 1988).
45. The episode is described in Steven D. Levitt and Stephen J. Dubner, *Freakonomics: A Rogue Economist Explores the Hidden Side of Everything* (New York: William Morrow, 2005), ch. 2. While Levitt cares about the power of information, I care about the power of humor.

46. Franz Rosenthal, *Humor in Early Islam* (Leiden: E. J. Brill, 1956); Khalid Kishtainy, *Arab Political Humor* (London: Quartet Books, 1985).
47. Kishtainy, *Arab Political Humor*, p. 18. The record on Muhammad's own attitude toward humor in religious matters is mixed at best; see, for example, Robert Spencer, *Islam Unveiled: Disturbing Questions About the World's Fastest Growing Faith* (New York: Encounter Books, 2003).
48. Rosenthal (*Humor in Early Islam*, pp. 147–48) provides a full list of "fools, wits, and entertainers," including all those mentioned here.
49. Frank Hessenland, "Sharif Kanaana on the Palestinian Sense of Humor," Dialogue with the Islamic World, April 2, 2005, available at: http://www.qantara.de/webcom/show_article.php/_c-478/_nr-235/i.html.
50. See Abu Hayyan al-Tawhidi, cited in Kishtainy, *Arab Political Humor*, p. 5.
51. It is the comic, after all, who breaks the bond of honor that ties the fool to his social obligations. This is explored in detail in William Ian Miller, *Humiliation: And Other Essays on Honor, Social Discomfort, and Violence* (Ithaca, N.Y.: Cornell University Press, 1993), ch. 4. See also chapter 8 in this volume.

INDEX